Birkhäuser

Trends in Mathematics

Research Perspectives Ghent Analysis and PDE Center

Volume 14

Series Editor

Michael Ruzhansky, Department of Mathematics, Ghent University, Gent, Belgium

Research Perspectives Ghent Analysis and PDE Center is a book series devoted to the publication of extended abstracts of seminars, conferences, workshops, and other scientific events related to the Ghent Analysis and PDE Center. The extended abstracts are published in the subseries Research Perspectives Ghent Analysis and PDE Center within the book series Trends in Mathematics. All contributions undergo a peer-review process to meet the highest standard of scientific literature.

Volumes in the subseries will include a collection of revised written versions of the communications or short research announcements or summaries, grouped by events or by topics. Contributing authors to the extended abstracts volumes remain free to use their own material as in these publications for other purposes (for example a revised and enlarged paper) without prior consent from the publisher, provided it is not identical in form and content with the original publication and provided the original source is appropriately credited.

Pamela E. Harris • Jaqueline G. Mesquita •
Yuriko Pitones
Editors

Women in Mathematics in Latin America

ICMAM 2024

Birkhäuser

Editors
Pamela E. Harris
Department of Mathematical Sciences
University of Wisconsin–Milwaukee
Milwaukee, WI, USA

Jaqueline G. Mesquita
Departamento de Matemática
Universidade de Brasília
Brasília, Brazil

Yuriko Pitones
Departamento de Matemáticas
Universidad Autónoma Metropolitana
Ciudad de México, Distrito Federal, Mexico

ISSN 2297-0215 ISSN 2297-024X (electronic)
Trends in Mathematics
ISSN 2948-1724 ISSN 2948-1732 (electronic)
Research Perspectives Ghent Analysis and PDE Center
ISBN 978-3-031-97856-2 ISBN 978-3-031-97857-9 (eBook)
https://doi.org/10.1007/978-3-031-97857-9

This book is published under the imprint Birkhäuser, www.birkhauser-science.com by the registered company Springer Nature Switzerland AG
The registered company address is: Gewerbestrasse 11, 6330 Cham, Switzerland

To all the women who paved the way, to those who walk it, and to those who will follow.

A todas las mujeres que trazaron el camino, a las que lo recorren y a las que vendrán.

A todas as mulheres que abriram caminho, às que o percorrem e às que ainda virão.

Preface

Mathematics has been, throughout history, a fundamental pillar of knowledge for humanity, which has influenced the development of science, technology, and society. However, women's participation in this field has been historically limited by multiple structural, social, and cultural barriers. For instance, the Fields Medal, often regarded as the equivalent of the Nobel Prize in Mathematics, was first awarded in 1936. Since then, only two women have received this prestigious honor, compared to 62 men. A similar pattern of underrepresentation is evident in the Abel Prize, established in 2002, where only one woman has been awarded to date. These disparities highlight the urgent need for change. Ensuring greater diversity is not just a matter of equity, it is essential for fostering innovation and advancing scientific progress.

In Latin America, these barriers are even more pronounced, often causing the contributions of women mathematicians to go unnoticed. Recently, collectives have emerged across Latin America to address the underrepresentation of women in mathematics. Many of these initiatives are affiliated with national Mathematical Societies and *Unión Matemática de América Latina y el Caribe* (UMALCA). However, discussions on this issue within Latin America are still relatively recent, and despite ongoing efforts, women remain significantly underrepresented in the field of mathematics, particularly in leadership positions. A striking example of this is that UMALCA only recently appointed its first female president, after a long history of exclusively male predecessors. This highlights the persistent challenges women face in reaching prominent and recognized positions within the mathematical community.

The book *Women in Mathematics Latin America* presents 19 chapters that bring together contributions from women mathematicians from Argentina, Brazil, Chile, Colombia, Costa Rica, Honduras, and Mexico. These researchers have developed studies in a wide variety of mathematical fields, ranging from algebra and geometry to analysis and statistics. Moreover, their work includes both theoretical and practical applications, addressing concrete problems in industry, society, and technology.

Each chapter reflects the rigor and creativity with which the authors and their collaborators approach their areas of expertise, showcasing new perspectives and innovative methods for solving mathematical problems. These contributions not only expand the horizons of mathematical knowledge but also highlight the impact and significance of women-led research within the Latin American mathematical community.

This book arises from the need to highlight and showcase the work of Latin American women mathematicians, focusing on their contributions in various areas of research. Through this collection of specialized articles, we aim to spread and recognize the work of the authors within the mathematical and scientific community, in line with global efforts to promote gender equity in science and technology. In this way, our goal is to support and amplify the voices of women mathematicians, fostering a more inclusive and diverse community. Increasing the visibility of their contributions not only acknowledges their achievements but also plays a key role in creating role models who can inspire future generations. By seeing women lead innovative projects, develop novel theories, and tackle complex mathematical challenges, we challenge stereotypes and encourage more girls and young women to pursue careers in mathematics.

Furthermore, this is the second volume in a series of publications resulting from initiatives under the framework of *ICMAM Latin America*, an organization committed to creating spaces that promote the integration of the Latin American mathematical community and its connection with other regions of the world. Through these pages, we reaffirm our commitment to promoting diversity in mathematics and recognizing the contributions of women in this field.

The editors would like to express their sincere gratitude to *ICMAM Latin America* for promoting initiatives that highlight the work of women mathematicians and for their ongoing support throughout the entire editing process of this book. We also acknowledge their valuable contribution in including events such as the *ISAAC-ICMAM Conference of Women in Mathematics 2024*, organized by Karina Navarro (Weizmann Institute of Science), Vanesa Hurtado (RPTU Kaiserslautern), and Milton Aguirre (Universidade de São Paulo), as well as the *Visibility of Latin American Women Mathematicians Project IMU-CWM 2024*, led by Yuriko Pitones (UAM), which have been crucial in raising the visibility and strengthening the mathematical community in the region.

We would like to express our sincere gratitude to Kunda Kambaso from Birkhäuser-Springer for her invaluable support. We extend our deepest appreciation to the Editor-in-Chief, Michael Ruzhansky (Ghent University), whose backing was crucial in making this editorial project possible. Additionally, we wish to acknowledge Duván Cardona (Ghent University), Scientific Director of *ICMAM Latin America* and a member of the Editorial Board, for his leadership and guidance in the execution of this project.

Finally, we would like to express our sincere gratitude to all the invited authors and co-authors who, through their valuable collaborative work, have formed the core of this book. This joint effort highlights the importance of teamwork in the construction of knowledge and the advancement of mathematics.

At the end of the preface, the photos of the authors and their collaborators are included, as it is essential for us to humanize and give a face to the ideas, bringing visibility to the people behind the studies and contributions presented. This gesture not only allows us to recognize their work but also fosters a more inclusive and diverse environment in our discipline.

We hope that this volume will not only serve as a valuable resource for the mathematical community but also as a source of inspiration for all those seeking to advance in this area of knowledge.

Milwaukee, WI, USA — Pamela E. Harris
Brasília, Brazil — Jaqueline G. Mesquita
Ciudad de México, Mexico — Yuriko Pitones
March 2025

Prefacio

Las matemáticas han sido, a lo largo de la historia, un pilar fundamental del conocimiento para la humanidad, influyendo en el desarrollo de la ciencia, la tecnología y la sociedad. Sin embargo, la participación de las mujeres en este campo ha sido históricamente limitada por múltiples barreras estructurales, sociales y culturales. Un ejemplo claro de esta desigualdad es la Medalla Fields, considerada el equivalente al Premio Nobel en Matemáticas, que se otorgó por primera vez en 1936. Desde entonces, solo dos mujeres han recibido este prestigioso reconocimiento, en comparación con 62 hombres. Un patrón similar de subrepresentación se observa en el Premio Abel, establecido en 2002, donde hasta la fecha solo una mujer ha sido galardonada. Estas disparidades destacan la necesidad urgente de un cambio. Garantizar una mayor diversidad no es solo una cuestión de equidad, sino que es esencial para fomentar la innovación y el progreso científico.

En América Latina, estas barreras son aún más pronunciadas, lo que ha provocado que las contribuciones de las mujeres matemáticas a menudo pasen desapercibidas. Recientemente, han surgido colectivos en toda América Latina con el objetivo de abordar la subrepresentación de las mujeres en matemáticas. Muchas de estas iniciativas están afiliadas a sociedades matemáticas nacionales y a la *Unión Matemática de América Latina y el Caribe* (UMALCA). No obstante, las discusiones sobre este tema en América Latina son relativamente recientes y, a pesar de los esfuerzos en curso, las mujeres siguen estando significativamente subrepresentadas en el campo de las matemáticas, particularmente en posiciones de liderazgo. Un ejemplo notable de esto es que UMALCA recientemente nombró a su primera presidenta, después de una larga historia de predecesores exclusivamente masculinos. Esto resalta los desafíos persistentes que enfrentan las mujeres para alcanzar posiciones prominentes y reconocidas dentro de la comunidad matemática.

El libro *Women in Mathematics Latin America* presenta 19 capítulos que reúnen contribuciones de mujeres matemáticas de Argentina, Brasil, Chile, Colombia, Costa Rica, Honduras y México. Estas investigadoras han desarrollado estudios en una amplia variedad de áreas matemáticas, desde álgebra y geometría hasta análisis y estadística. Además, su trabajo incluye tanto aplicaciones teóricas como prácticas, abordando problemas concretos en la industria, la sociedad y la tecnología.

Cada capítulo refleja el rigor y la creatividad con los que las autoras y sus colaboradores abordan sus áreas de especialización, mostrando nuevas perspectivas y métodos innovadores para resolver problemas matemáticos. Estas contribuciones no solo amplían los horizontes del conocimiento matemático, sino que también resaltan el impacto y la importancia de la investigación liderada por mujeres dentro de la comunidad matemática latinoamericana.

Este libro surge de la necesidad de visibilizar y destacar el trabajo de las matemáticas latinoamericanas, poniendo énfasis en sus contribuciones en diversas áreas de investigación. A través de esta recopilación de artículos especializados, buscamos difundir y reconocer el trabajo de las autoras y sus colaboradores dentro de la comunidad matemática y científica, en línea con los esfuerzos globales por promover la equidad de género en la ciencia y la tecnología. De esta manera, nuestro objetivo es apoyar y amplificar las voces de las mujeres matemáticas, fomentando una comunidad más inclusiva y diversa. Aumentar la visibilidad de sus contribuciones no solo reconoce sus logros, sino que también desempeña un papel clave en la creación de modelos a seguir que puedan inspirar a futuras generaciones. Al ver a mujeres liderar proyectos innovadores, desarrollar nuevas teorías y abordar problemas matemáticos complejos, desafiamos estereotipos y motivamos a más niñas y jóvenes a seguir carreras en matemáticas.

Además, este es el segundo volumen de una serie de publicaciones derivadas de iniciativas dentro del marco de *ICMAM Latin America*, una organización comprometida con la creación de espacios que promuevan la integración de la comunidad matemática latinoamericana y su conexión con otras regiones del mundo. A través de estas páginas, reafirmamos nuestro compromiso de promover la diversidad en matemáticas y reconocer las contribuciones de las mujeres en este campo.

Las editoras desean expresar su sincero agradecimiento a *ICMAM Latin America* por impulsar iniciativas que destacan el trabajo de las mujeres matemáticas y por su continuo apoyo a lo largo de todo el proceso de edición de este libro. También reconocemos su valiosa contribución al incluir eventos como la *ISAAC-ICMAM Conference of Women in Mathematics 2024*, organizada por Karina Navarro (Weizmann Institute of Science), Vanesa Hurtado (RPTU Kaiserslautern) y Milton Aguirre (Universidade de São Paulo), así como el proyecto *Visibility of Latin American Women Mathematicians IMU-CWM 2024*, liderado por Yuriko Pitones (UAM), los cuales han sido cruciales para aumentar la visibilidad y fortalecer la comunidad matemática en la región.

Queremos expresar nuestro más sincero agradecimiento a Kunda Kambaso de Birkhäuser-Springer por su invaluable apoyo. Extendemos nuestro profundo reconocimiento al editor en jefe, Michael Ruzhansky (Ghent University), cuyo respaldo fue fundamental para hacer posible este proyecto editorial. Además, deseamos reconocer a Duván Cardona (Ghent University), director científico de *ICMAM Latin America* y miembro del comité editorial, por su liderazgo y guía en la ejecución de este proyecto.

Finalmente, queremos expresar nuestro más sincero agradecimiento a todas las autoras invitadas, coautoras y coautores, quienes, a través de su valioso trabajo colaborativo, han conformado el núcleo de este libro. Este esfuerzo conjunto resalta la importancia del trabajo en equipo en la construcción del conocimiento y el avance de las matemáticas.

Al final del prefacio, se incluyen las fotografías de las autoras y sus colaboradores, ya que para nosotras es esencial humanizar y dar un rostro a las ideas, brindando visibilidad a las personas detrás de los estudios y contribuciones presentadas. Este gesto no solo nos permite reconocer su trabajo, sino que también fomenta un entorno más inclusivo y diverso en nuestra disciplina.

Esperamos que este volumen no solo sea un recurso valioso para la comunidad matemática, sino también una fuente de inspiración para todas aquellas personas que buscan avanzar en esta área del conocimiento.

Milwaukee, WI, USA
Brasília, Brazil
Ciudad de México, Mexico
Marzo 2025

Pamela E. Harris
Jaqueline G. Mesquita
Yuriko Pitones

Prefácio

A matemática tem sido, ao longo da história, um pilar fundamental do conhecimento para a humanidade, influenciando o desenvolvimento da ciência, da tecnologia e da sociedade. No entanto, a participação das mulheres neste campo tem sido historicamente subrepresentada devido às múltiplas barreiras estruturais, sociais e culturais. Um exemplo claro dessa desigualdade é a *Medalha Fields*, considerada equivalente ao Prêmio Nobel em Matemática, que foi concedida pela primeira vez em 1936. Desde então, apenas duas mulheres receberam este prestigiado reconhecimento, em comparação com 62 homens. Um padrão semelhante de subrepresentação é observado no *Prêmio Abel*, estabelecido em 2002, onde até hoje apenas uma mulher foi premiada. Essas disparidades destacam a necessidade urgente de mudança. Garantir uma maior diversidade não é apenas uma questão de equidade, mas é essencial para fomentar a inovação e o progresso científico.

Na América Latina, essas barreiras são ainda mais evidentes, o que fez com que as contribuições das mulheres matemáticas frequentemente passassem despercebidas. Recentemente, surgiram coletivos em toda a América Latina com o objetivo de discutir a subrepresentatividade das mulheres na matemática. Muitas dessas iniciativas estão ligadas às sociedades matemáticas nacionais e à *União Matemática da América Latina e do Caribe* (UMALCA). No entanto, as discussões sobre este tema na América Latina são relativamente recentes e, apesar dos esforços em andamento, as mulheres ainda estão significativamente subrepresentadas no campo da matemática, particularmente em posições de liderança. Um exemplo notável disso é que a UMALCA nomeou recentemente sua primeira presidente, após uma longa história de predecessores exclusivamente masculinos. Isso destaca os desafios persistentes enfrentados pelas mulheres para alcançar posições de destaque e de prestígio dentro da comunidade matemática.

O livro *Women in Mathematics Latin America* apresenta 19 capítulos que reúnem contribuições de mulheres matemáticas da Argentina, Brasil, Chile, Colômbia, Costa Rica, Honduras e México. Essas pesquisadoras desenvolveram estudos em uma ampla variedade de áreas matemáticas, desde álgebra e geometria até análise e estatística. Além disso, seus trabalhos incluem tanto aplicações teóricas quanto práticas, abordando problemas concretos na indústria, sociedade e tecnologia.

Cada capítulo traz o rigor e a criatividade com os quais as autoras e seus colaboradores discutem suas áreas de especialização, mostrando novas perspectivas e métodos inovadores para resolver certos problemas matemáticos. Essas contribuições não apenas ampliam os horizontes do conhecimento matemático, mas também ressaltam o impacto e a importância da pesquisa liderada por mulheres dentro da comunidade matemática latino-americana.

Este livro surge da necessidade de visibilizar e destacar o trabalho das matemáticas latino-americanas, enfatizando suas contribuições em diversas áreas de pesquisa. Através desta coletânea de artigos especializados, buscamos divulgar e reconhecer o trabalho das autoras e seus colaboradores dentro da comunidade matemática e científica, se unindo aos esforços globais para promover a equidade de gênero na ciência e tecnologia. Dessa forma, nosso objetivo é apoiar e amplificar as vozes das mulheres matemáticas, promovendo uma comunidade mais inclusiva e diversa. Aumentar a visibilidade de suas contribuições não só reconhece seus feitos, mas também desempenha um papel crucial na criação de modelos a seguir que possam inspirar futuras gerações. Ao ver mulheres liderando projetos inovadores, desenvolvendo novas teorias e abordando problemas matemáticos complexos, desafiamos estereótipos e motivamos mais meninas e jovens a seguir carreiras em matemática.

Além disso, este é o segundo volume de uma série de publicações oriundas de iniciativas dentro do marco da *ICMAM Latin America*, uma organização comprometida com a criação de espaços que promovam a integração da comunidade matemática latino-americana e sua conexão com outras regiões do mundo. Por meio deste livro, reafirmamos nosso compromisso em promover a diversidade em matemática e reconhecer as contribuições das mulheres neste campo.

As editoras desejam expressar seu sincero agradecimento à *ICMAM Latin America* por apoiar iniciativas que destacam o trabalho das mulheres matemáticas e por seu contínuo apoio ao longo de todo o processo de edição deste livro. Também reconhecemos sua valiosa contribuição ao incluir eventos como a *ISAAC-ICMAM Conference of Women in Mathematics 2024*, organizada por Karina Navarro (Weizmann Institute of Science), Vanessa Hurtado (RPTU Kaiserslautern) e Milton Aguirre (Universidade de São Paulo), assim como o projeto *Visibility of Latin American Women Mathematicians IMU-CWM 2024*, liderado por Yuriko Pitones (UAM), os quais foram cruciais para aumentar a visibilidade e fortalecer a comunidade matemática na região.

Queremos expressar nosso mais sincero agradecimento a Kunda Kambaso da Birkhäuser-Springer por seu apoio inestimável. Estendemos nosso profundo reconhecimento ao editor-chefe, Michael Ruzhansky (Ghent University), cujo apoio foi fundamental para viabilizar este projeto editorial. Além disso, desejamos agradecer Duván Cardona (Ghent University), diretor científico da *ICMAM Latin America* e membro do comitê editorial, por sua liderança e orientação na execução deste projeto.

Finalmente, queremos expressar nosso mais sincero agradecimento a todas as autoras convidadas, coautoras e coautores, que, através de seu valioso trabalho colaborativo, formaram o núcleo deste livro. Esse esforço coletivo destaca a importância do trabalho em equipe na construção do conhecimento e no avanço da matemática.

Ao final do prefácio, são incluídas as fotografias das autoras e de seus colaboradores, pois para nós é essencial humanizar e dar um rosto às ideias, dando visibilidade às pessoas por trás dos estudos e contribuições apresentadas. Este gesto não apenas nos permite reconhecer seu trabalho, mas também fomentar um ambiente mais inclusivo e diverso em nossa disciplina.

Esperamos que este volume não seja apenas um recurso valioso para a comunidade matemática, mas também uma fonte de inspiração para todas as pessoas que buscam avançar nesta área do conhecimento.

Milwaukee, WI, USA
Brasília, Brazil
Ciudad de México, Mexico
Março de 2025

Pamela E. Harris
Jaqueline G. Mesquita
Yuriko Pitones

Contents

Contributing Authors and Editors

Martha Álvarez-Ramírez

Gabriela Araujo-Pardo

Verónica Arredondo

Leticia Ramírez

Valera Falcon

Diana Bueno-Carreño

Mario E. Duarte

Mauricio Frieri

Leasly Campa-Raymundo

Andrea Ceretani

Cuiyu He

Lin Mu

Weiwei Hu

Carlos N. Rautenberg

María Isabel Cortez

Juliana Fernandes

Julia do Prado Pazzini

Juliana Honda Lopes

Myrian Gonzalez Orellana

Jose M. Alvarenga Rodriguez

Amanda Montejano

Samaria Montenegro

Carolina Mosquera

Rocio Balderrama

Daniela Paiva

Victoria Paternostro

Emily King

Rocio Nores

Emily Quesada-Herrera

Lesky Rivas Martínez

Mariel Sáez Trumper

Maria Fernanda Espinal

Adriana Sánchez Chavarría

Maya Stein

Pamela E. Harris

Jaqueline G. Mesquita

Yuriko Pitones

Contributors

José Mauricio Alvarenga Rodríguez Master's Program in Mathematics, Universidad Nacional Autónoma de Honduras, Tegucigalpa, Honduras

Martha Alvarez Ramírez Department of Mathematics, Universidad Autónoma Metropolitana, México City, Mexico

Gabriela Araujo-Pardo Instituto de Matemáticas-Campus Juriquilla, Universidad Nacional Autónoma de México, Juriquilla, Mexico

Verónica Arredondo Unidad Académica de Matemáticas, Universidad Autónoma de Zacatecas, Zacatecas, Mexico

Rocío Balderrama Universidad de Buenos Aires e IMAS-CONICET, Buenos Aires, Argentina

Diana H. Bueno-Carreño Pontificia Universidad Javeriana seccional Cali, Cali, Colombia

Leasly A. Campa-Raymundo Department of Mathematics, Universidad Autónoma Metropolitana, Mexico City, Mexico

Andrea Ceretani Universidad de Buenos Aires (UBA) e Instituto de Investigaciones Matemáticas Luis A. Santaló (IMAS), UBA-CONICET, Buenos Aires, Argentina

María Isabel Cortez Pontificia Universidad Católica de Chile, Macul, Chile

Mario E. Duarte DyG Representaciones SAS, Bogotá, Colombia

Maria Fernanda Espinal CMM-Center for Mathematical Modeling, Santiago, Chile

Brenda Valeria López Falcón Instituto Nacional de Geografía e Historia (INEGI), Aguascalientes, Mexico

Juliana Fernandes da Silva Instituto de Matemática da Universidade Federal do Rio de Janeiro, Rio de Janeiro, Brazil

Mauricio Frieri Universiteit Leiden, Leiden, Netherlands

Myrian Sadith González Orellana School of Mathematics and Computer Science, Universidad Nacional Autónoma de Honduras, Tegucigalpa, Honduras

Cuiyu He Department of Mathematics, University of Georgia, Athens, GA, USA

Weiwei Hu Department of Mathematics, University of Georgia, Athens, GA, USA

Emily J. King Mathematics Department, Colorado State University, Fort Collins, CO, USA

Juliana Honda Lopes Instituto de Matemática e Estatística, Universidade de São Paulo, São Paulo, Brazil

Amanda Montejano UMDI, Facultad de Ciencias, UNAM Juriquilla, Juriquilla, Mexico

Samaria Montenegro Universidad de Costa Rica, Escuela de Matemática-CIMPA, San José, Costa Rica

Carolina Mosquera Departamento de Matemática, Universidad de Buenos Aires, Buenos Aires, Argentina

Lin Mu Department of Mathematics, University of Georgia, Athens, GA, USA

Rocío Nores Universidad de Buenos Aires, Instituto de Matemática Luis Santaló (IMAS-CONICET-UBA), Buenos Aires, Argentina

Daniela Paiva Instituto de Matemática Pura e Aplicada, Rio de Janeiro, Brazil

Victoria Paternostro Universidad de Buenos Aires, Instituto de Matemática Luis Santaló (IMAS-CONICET-UBA), Buenos Aires, Argentina

Julia do Prado Pazzini Instituto de Matemática da Universidade Federal do Rio de Janeiro, Rio de Janeiro, Brazil

Emily Quesada-Herrera University of Lethbridge, Lethbridge, AB, Canada

Leticia Adriana Ramírez Hernández Unidad Académica de Matemáticas, Universidad Autónoma de Zacatecas, Zacatecas, Mexico

Carlos N. Rautenberg Department of Mathematical Sciences and Center for Mathematics and Artificial Intelligence (CMAI), George Mason University, Fairfax, VA, USA

Lesky Ibeth Rivas Martínez Universidad Nacional Autónoma de Honduras, Tegucigalpa, Honduras

Adriana Cristina Sánchez Chavarría Universidad de Costa Rica, San Jose, Costa Rica

Maya Stein University of Chile, Santiago, Chile

Mariel Sáez Trumper Pontificia Universidad Católica de Chile, Santiago, Chile

A Quick Dive into Celestial Mechanics

Martha Alvarez Ramírez

Abstract Celestial mechanics applies classical mechanics, combining mathematics and physics to explain the motion of celestial bodies under various forces. One of the most renowned problems in this field is the n-body problem, which seeks to describe the motion of n bodies under their mutual gravitational attraction. In this work, we present a brief introduction to celestial mechanics and the n-body problem, along with some historical background and key results.

Keywords n-Body problem · Central configurations · Stability

1 Introduction

Throughout human history, various cultures have attempted to interpret and explain the motion of objects. However, it was not until the seventeenth century that a true understanding of the causes of motion emerged. Inspired by Galileo's work on falling bodies, Kepler's laws of planetary motion, and the movements of celestial bodies, Isaac Newton (1643–1727) formulated the law of universal gravitation, which explains why planets follow predictable and calculable orbits.

Newton built on the groundbreaking ideas of Copernicus, Galileo, and Kepler, unifying them into a coherent framework. His mechanics, particularly the three laws of motion, enabled the precise prediction of an object's movement based on the principle of gravity. This marked the beginning of modern astronomy, where advances were accumulated in both the mathematical field (celestial mechanics) and the physical sciences (Astrophysics). In mathematical terms, the goal of celestial mechanics is to study a system of differential equations known as the n-body problem, which aims to describe the motion of n point of positive masses in three-

M. Alvarez Ramírez (✉)
Department of Mathematics, Universidad Autónoma Metropolitana - Iztapalapa, México City, Mexico
e-mail: mar@xanum.uam.mx

P. E. Harris et al. (eds.), *Women in Mathematics in Latin America*,
Research Perspectives Ghent Analysis and PDE Center 14,
https://doi.org/10.1007/978-3-031-97857-9_1

dimensional Euclidean space, governed by Newton's laws of gravitation. We should note that, from a physical point of view, this is an idealized model, as celestial bodies are not point masses.

Newton's research focuses on the motion of the Moon, which, to a first approximation, involves three bodies: the Sun, the Earth, and the Moon. However, his struggle to solve the three-body problem led him to remark to astronomer John Machin that he had "never had a headache except in his studies of the Moon", see [23]. This challenge consolidated the three-body problem as one of the most famous in mathematics.

A long-standing question that has always troubled humanity is whether the planetary system will maintain its current form or if, in the future, one of the planets might escape the system—or, perhaps worse, collide with another celestial body. In modern terms, the question is: *Is the solar system stable or unstable?* Newton concluded that divine intervention was necessary to preserve the equilibrium of the solar system. In the late nineteenth century, King Oscar II of Sweden, advised by Mittag-Leffler, established a contest to celebrate his 60th birthday, offering a prize to anyone who could solve four mathematical problems from various fields of research. One of these problems was the n-body problem in celestial mechanics, which is essential for understanding the stability of the solar system. Henri Poincaré emerged as the winner with his memoir on the three-body problem, published in the journal *Acta Mathematica* in 1890. This work made a major contribution to the development of mathematics, especially in advancing the understanding of the three-body problem. In particular, his methods for analyzing stability led to the development of topology and the qualitative theory of differential equations, while also contributing to the emergence of chaos theory.

Since the publication of Newton's *Philosophiæ Naturalis Principia Mathematica*, hundreds of papers have been written on the n-body problem, exploring special solutions, asymptotic estimates, collision dynamics, the existence or absence of prime integrals, series solutions, and collisionless singularities, among other topics. Despite these extensive contributions, we are still far from fully understanding the overall dynamics of all solutions. The two-body problem, also known as Kepler's problem, has been fully solved. However, for $n \geq 3$, the n-body problem remain an open problem.

This article proceeds as follows: In Sect. 2, we describe the two-body problem, which reduces to the Kepler problem. In this formulation, the motion of one particle is governed by the gravitational influence of a second particle fixed at the origin. This reduction simplifies the dynamics into a system of ordinary differential equations, allowing the two-body problem to be fully solvable. In Sect. 3; next, in Sect. 4, we introduce the n-body problem from the point of view of Hamiltonian systems. Sect. 5 introduces the restricted n-body problems, focusing particularly on the most widely studied case: the planar circular restricted three-body problem (CRTBP). The final section of the paper highlights some of my contributions to advancing the understanding of the dynamics of the n-body problem.

2 The Beginnings

Before addressing the general n-body problem, it is essential to first examine the two-body problem. The Kepler problem focuses on determining the motion of two bodies that orbit each other due to their mutual gravitational attraction, as described by Kepler's laws. Newton's work in the *Principia Mathematica* offered a groundbreaking solution to this problem by deriving Kepler's laws from his own law of universal gravitation combined with the laws of motion.

To formulate the Kepler problem, we begin by recalling Newton's law of universal gravitation. This law states that every mass in the universe attracts every other mass with a force F that is directly proportional to the product of their masses and inversely proportional to the square of the distance d between their centers. Mathematically, it is expressed as:

$$F = G\frac{m_1 m_2}{d^2},$$

where $G = 6.67260 \times 10^{-11}\,\mathrm{Nm^2\,kg^{-2}}$ is the *universal gravitational constant.*

Consider the motion of two masses, m_1 and m_2, with position vectors $\mathbf{r}_1$ and $\mathbf{r}_2$ relative to a fixed origin in an inertial frame. According to Newton's second law and the law of universal gravitation, the equation of motion becomes:

$$m_1\ddot{\mathbf{r}}_1 = G\frac{m_1 m_2}{r^3}\mathbf{r}, \quad \text{and} \quad m_2\ddot{\mathbf{r}}_2 = -G\frac{m_1 m_2}{r^3}\mathbf{r} \tag{1}$$

where $\mathbf{r} = \mathbf{r}_2 - \mathbf{r}_1$ and $r = \|\mathbf{r}\|$. Now, we focus on the motion of m_2 relative to m_1. Writing $\ddot{\mathbf{r}} = \ddot{\mathbf{r}}_2 - \ddot{\mathbf{r}}_1$, and using Eq. (1), we obtain

$$\frac{d^2\mathbf{r}}{dt^2} + \mu\frac{\mathbf{r}}{r^3} = 0, \tag{2}$$

where $\mu = G(m_1 + m_2)$. It is verified that the angular momentum vector $\mathbf{c} = \mathbf{r} \times \mathbf{v}$ is conserved, where $\mathbf{v}$ is the velocity vector of the position $\mathbf{r}$. If $\mathbf{c} \neq \mathbf{0}$, the motion occurs in the plane orthogonal to $\mathbf{c}$. This allows the use of polar coordinates, namely, $\mathbf{r} = \langle r\cos\theta, r\sin\theta\rangle$. Next, we consider Kepler's second law, which states that $\frac{dA}{dt} = \frac{1}{2}r^2\frac{d\theta}{dt}$, representing the area swept out by the position vector per unit of time, see [22]. Thus, we obtain a scalar equation for the relative motion in the plane (r, θ):

$$\ddot{r} - r\dot{\theta}^2 = -\frac{\mu}{r^2}.$$

To solve this equation and express r as a function of θ, we first make the substitution $u = 1/r$. To eliminate time, we use the constant of motion $c = \|\mathbf{c}\| = r^2\dot{\theta}$. Applying these changes leads us to the equation:

$$\frac{d^2u}{d\theta^2} + u = \frac{\mu}{c^2}.$$

By solving this equation and considering $r = 1/u$, we derive the equation of the trajectory of m_2 in its motion relative to m_1:

$$r = \frac{c^2/\mu}{1 + e\cos f}.$$

This represents the general equation of a conic section in polar coordinates, with one focus located at the origin. Here, e is a constant of integration representing the eccentricity, $f = \theta - \omega$, with ω being a phase angle relative to a reference direction.

The four possible types of conics are as follows: (1) when $e = 0$, the orbit is a circle; (2) for $0 < e < 1$, the orbit is an ellipse; (3) at $e = 1$, the orbit is a parabola, and (4) for $e > 1$, the orbit is a hyperbola. In our solar system, all the planets have $e < 1$, which means they follow elliptical orbits around the Sun. For instance, Mercury has an eccentricity of $e = 0.21$, Earth has $e = 0.017$, Pluto has $e = 0.25$, and Saturn has $e = 0.052$; see, for example, [22].

3 The n-Body Problem

The Newtonian n-body problem describes the dynamics of n point particles of positive masses m_i located at $\mathbf{q}_i \in \mathbb{R}^d$, moving according to the Newton's laws under their mutual gravitational forces. The equations of motion of this n-body problem are as follows:

$$m_i\ddot{\mathbf{q}}_i = -\sum_{j=1, j\neq i}^{n} Gm_i m_j \frac{\mathbf{q}_i - \mathbf{q}_j}{r_{ij}^3}, \qquad 1 \leq i \leq n, \tag{3}$$

where the dot represents differentiation with respect to time $\frac{d}{dt}$, G is the gravitational constant, $r_{ij} = \|\mathbf{q}_i - \mathbf{q}_j\|$ is the mutual distance between the masses m_i and m_j and $\|\cdot\|$ denotes the Euclidean norm. Without loss of generality, we can choose the time unit such that $G = 1$. If $d = 2$, the problem is planar (in $\mathbb{R}^2$), while $d = 3$ corresponds to the spatial case (in $\mathbb{R}^3$). For further details, see the book by Wintner [28]. We stress that system (3) does not have any equilibrium solutions. For more details see [18], page 73.

The system of equations (3) with $d = 3$ has ten first integrals: the linear momentum vector is constant, indicating that the center of mass of the system moves in a straight line at a constant velocity, contributing six integrals. Additionally, there are three components of angular momentum and one for energy. These constants of motion reduce the dimension of the phase space to $6n - 10$ (see [18]). In 1887, Bruns demonstrated that the equations of motion for the n-body problem (with $n > 2$) admit no integrals algebraic in time, position, and velocity, other than the ten classical integrals. Nevertheless, Bruns' proof contained an error, which Poincaré pointed out in 1896, along with the appropriate correction he provided in [24]. Since the problem for $n > 2$ cannot be solved by quadratures, the community has sought particular solutions, such as homographic ones (defined below).

The configuration space or space of positions is defined by

$$C = \{\mathbf{q} = (\mathbf{q}_1, \ldots, \mathbf{q}_n) \in (\mathbb{R}^d)^n : \mathbf{q}_i \neq \mathbf{q}_j \quad i \neq j\}.$$

The configuration $\mathbf{q} = (\mathbf{q}_1, \ldots, \mathbf{q}_n)$ is referred to as *central* if the acceleration vector for each body is a common scalar multiple of its position vector relative to the center of mass. This indicates that there exists a positive constant λ such that

$$\ddot{\mathbf{q}}_i = \lambda(\mathbf{q}_i - \mathbf{c}_m), \qquad i = 1, \ldots, n, \tag{4}$$

where $\mathbf{c}_m = \dfrac{1}{M}(m_1\mathbf{q}_1 + \cdots + m_n\mathbf{q}_n)$ and $M = m_1 + \cdots + m_n$ is the center of mass of the n bodies and the total mass, respectively. Thus the given configuration $(\mathbf{q}_1, \ldots, \mathbf{q}_n) \in C$ of the n-body problem with positive masses $m_1, \ldots, m_n$, is central if there exists a λ such that $(\lambda, \mathbf{q}_1, \ldots, \mathbf{q}_n)$ is a solution of the system

$$\sum_{j=1, j\neq i}^{n} m_j \frac{\mathbf{q}_i - \mathbf{q}_j}{r_{ij}^3} = \lambda(\mathbf{q}_i - \mathbf{c}_m), \qquad 1 \leq i \leq n. \tag{5}$$

The planar central configurations in which bodies maintain the same arrangement while rotating in a circular motion over time are referred to as *relative equilibria*. Also, we point out that central configurations serve as the initial conditions for homographic orbits in the n-body problem. Additionally, they play a crucial role in identifying bifurcation behavior related to the topology of the manifolds defined by fixed values of the angular momentum and energy integrals, see [25]. However, finding and classifying central configurations is generally a very difficult algebraic problem, and the problem remains largely unsolved. For more information on this topic, the reader is referred to [20].

We remark that two central configurations are considered equivalent if there exists a rotation about the center of mass of the bodies and a homothety that transforms one configuration into the other; see [20, 28]. This relationship is an equivalence relation; therefore, when we refer to central configurations, we are actually referring to classes of equivalent configurations.

The simplest solutions in the planar n-body problem are known as *homographic* solutions, where the configuration remains fixed, aside from rotations and homoth-

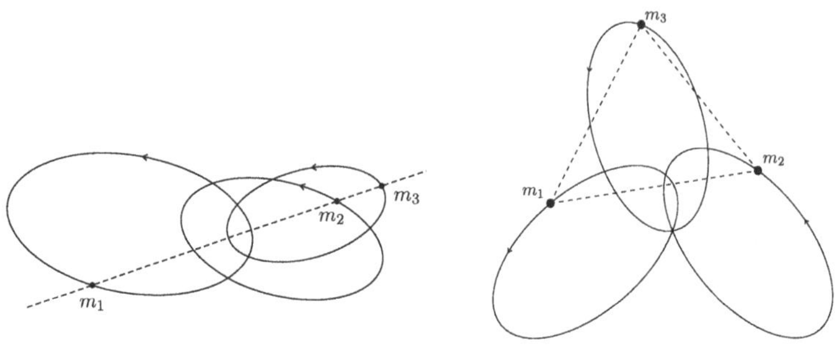

Fig. 1 The Eulerian and Lagrange homographic solutions of the three-body problem

eties (scalar multiplication). *Relative equilibria* are a specific type of homographic solution characterized by the system rotating around its center of mass $\mathbf{c}_m \in \mathbb{R}^2$ like a rigid body with a non-zero angular velocity different from zero.

The first periodic solutions to the three-body problem were homographic solutions. Specifically, in 1765, Euler [13] discovered those in which the three bodies remain aligned along a straight line, with their positions changing over time but always maintaining collinearity, each following a specific Keplerian orbit. Later, in 1772, Lagrange [17] described the homographic solutions where the three bodies form an equilateral triangle at any instant of the motion and at the same time each body travels along a specific Keplerian orbit about the center of masses of the system. They are shown in Fig. 1. It is worth noting that in a rotating system, homographic solutions correspond to fixed central configurations (see [20]). In summary, the three-body problem admits three collinear central configurations, one for each possible ordering of the masses along the line, as well as two equilateral triangle central configurations.

The question of finiteness of the central configurations was first formulated as a conjecture by Chazy in 1918 and later by Wintner in 1941. It was subsequently included by Smale as problem number six on his list of problems to be addressed in the twenty-first century in [26]. This can be stated as: *Given n bodies with positive masses $m_1, \ldots, m_n$, is the number of relative equilibria finite?*

It is worth noting that there are partial answers to the problem posed by Smale. For instance, given a fixed mass vector $m = (m_1, \ldots, m_n)$ and a specified ordering of the bodies along a line, Moulton [21] demonstrated that, up to translation and scaling, there exists a unique collinear central configuration. In total, there are exactly n!/2 distinct classes of collinear central configurations, each corresponding to a possible ordering of the bodies. Recently, Hampton and Moeckel [16] demonstrated that for any set of four masses, a finite number of central configuration classes exists. For five or more masses, this result remains unproven; however, Albouy and Kaloshin [1] have recently made significant contributions to the case of five masses. They show that there are only a finite number of equivalence classes of central configurations if the masses do not belong to a closed algebraic subset of codimension-2 of the mass space, that is, the case when the positive masses do not satisfy some explicit polynomial relations.

4 Hamiltonian Approach

In this section, we will show that the n-body problem can be equivalently using Hamiltonian formalism. In what follows, we will be considering $d = 3$, which naturally includes $d = 2$, as the two-dimensional case is a subset of the three-dimensional one.

The equations of motion (3) can then be expressed as a system of $6n$ first-order differential equations given by

$$\dot{\mathbf{q}} = M^{-1}\mathbf{p}, \qquad \dot{\mathbf{p}} = -\nabla U(\mathbf{q}), \tag{6}$$

where $\mathbf{p} = (\mathbf{p}_1, \mathbf{p}_2, \ldots, \mathbf{p}_n) \in \mathbb{R}^{3n}$ is the momentum vector defined by $\mathbf{p}_i = m_i \dot{\mathbf{q}}_i$, and

$$\frac{\partial}{\partial \mathbf{q}_i} = \left(\frac{\partial}{\partial x_i}, \frac{\partial}{\partial y_i}, \frac{\partial}{\partial z_i}\right)^T, \quad \nabla = \left(\frac{\partial}{\partial \mathbf{q}_1}, \ldots, \frac{\partial}{\partial \mathbf{q}_n}\right)^T,$$

with $M = \mathrm{diag}(m_1, m_1, m_1, \ldots, m_n, m_n, m_n)$ being the mass matrix and $\nabla U(\mathbf{q})$ the gradient vector of minus the potential energy

$$U(\mathbf{q}) = U(\mathbf{q}_1, \mathbf{q}_2, \ldots, \mathbf{q}_n) = \sum_{1 \le i < j \le n} \frac{m_i m_j}{\|\mathbf{q}_j - \mathbf{q}_i\|}.$$

On the other hand, Eqs. (3) are not defined in the set

$$\Delta = \bigcup_{1 \le i < j \le n} \Delta_{ij},$$

where $\Delta_{ij} = \{(\mathbf{q}_i, \mathbf{q}_j) \in \mathbb{R}^{3n} : \mathbf{q}_i = \mathbf{q}_j\}$, representing that collisions occur. The domain of definition for the system (6), or the *phase space*, is given by

$$\wp = \{(\mathbf{q}, \mathbf{p}) \in \mathbb{R}^{6n} : \mathbf{q} \in (\mathbb{R}^{3n} \setminus \Delta), \ \mathbf{p} \in \mathbb{R}^{3n}\}.$$

Thus, we can now define the configuration space as $C = \{\mathbf{q} \in \mathbb{R}^{3n} \setminus \Delta\}$.

The motion equations for the system (6) can be rewritten as the Hamiltonian system:

$$\dot{\mathbf{q}}_i = \frac{\mathbf{p}_i}{m_i} = \frac{\partial H}{\partial \mathbf{p}_i}, \qquad \dot{\mathbf{p}}_i = -\sum_{j \ne i} \frac{m_i m_j (\mathbf{q}_i - \mathbf{q}_j)}{\|\mathbf{q}_j - \mathbf{q}_i\|^3} = -\frac{\partial H}{\partial \mathbf{q}_i},$$

where the Hamiltonian function is given by

$$H(\mathbf{q}, \mathbf{p}) = \sum_{i=1}^{n} \frac{\|\mathbf{p}_i\|^2}{2m_i} + U(\mathbf{q}).$$

The vectors $\mathbf{q}$ and $\mathbf{p}$ are called *conjugate variables*; see, for example, [18]. In fact, $H(\mathbf{q}, \mathbf{p})$ represents the total energy of the system, where $K(\mathbf{p}) = \sum_{i=1}^{n} \frac{\|\mathbf{p}_i\|^2}{2m_i}$ is the kinetic energy, and $U(\mathbf{q})$ is the potential energy.

5 The Restricted n-Body Problem

The solar system offers limited applications of the general n-body problem. This results an intriguing scenario where a more complex general problem is less useful than a simpler formulation.

A *restricted n-body problem* involves describing the motion of an infinitesimal mass m_0 under the gravitational influence of $n-1$ positive masses $m_1, \ldots, m_{n-1}$, referred to as *primaries*. Since m_0 does not perturb the motion of the primaries, the latter move according to a solution γ of the $(n-1)$-body problem. In [2], Alvarez and Llibre analyzed various combinations of restricted three-body problems concerning the possible two-body motions of the primaries and the positions of the infinitesimal particle, enumerating 30 distinct problems.

The planar circular restricted three-body problem (CRTBP) is the most extensively studied restricted problem. In this scenario, two massive bodies, m_1 and m_2 (referred to as the primaries), move in circular orbits around their common center of mass. The mass of the third body, m_0, is negligible, allowing it to have little impact on the motion of the two more massive bodies. Additionally, the motion of m_0 is confined to the orbital plane of the primaries. This problem is typically analyzed in a rotating coordinate system where the positions of the primaries remain fixed. For more information on restricted problems, see Szebehely's book [27].

In this synodic reference frame, the CRTBP has five equilibrium points: L_1, L_2, and L_3, which are collinear points (called Euler points) along the line connecting the two massive bodies. The other two points, L_4 and L_5, known as the Lagrangian points, form the vertices of equilateral triangles, with the two large masses occupying the other two vertices. The common base of these triangles lies along the line connecting the center of mass of the two primary bodies. Indeed these points correspond to the relative equilibria in the sidereal reference frame, making them the simplest solutions to the three-body problem, moreover they are periodic. The stability of the equilibrium points L_i $(i = 1, \ldots, 5)$ is of significant interest. The three collinear points, L_1, L_2, and L_3, are always unstable. However, periodic orbits, known as Halo orbits, exist around these points [27], making them ideal parking orbits for applications like communication satellites and deep-space radio telescopes [14]. The stability of the Lagrange points L_4 and L_5 depends on the mass ratio between the two primary bodies. In Earth-Moon and Sun-Earth systems, these points are linearly stable [19], meaning that a spacecraft stationed there would require minimal fuel to maintain its position. On the other side, in systems like Pluto-Charon, points L_4 and L_5 are unstable [15].

6 Research Done in Celestial Mechanics

At this point, I would like to provide some aspect of my contributions to the understanding of the dynamics of the n-body problem, particularly for cases where $n = 3, 4$ and 5.

The main result from my PhD thesis, detailed in [3], addresses a special case of the isosceles plane problem. In this case, two point particles with equal masses ($m_1 = m_2 = 1/2$) move along a straight line L, while a third, infinitesimal particle moves along a line orthogonal to L that passes through the center of mass of m_1 and m_2. Parabolic orbital manifolds at both infinities are constructed. By employing geometric arguments and model symmetries, we establish the existence of two heteroclinic orbits connecting these infinities. Numerical analysis then confirms that the parabolic orbital manifolds intersect transversally along these heteroclinic orbits. Finally, we demonstrate the non-integrability of the system.

The article published in 2009, in collaboration with C. Vidal [8], has undoubtedly marked a turning point in my research career, as it has accumulated the most citations in my scientific work. n this paper, we present the study of the dynamics of the restricted equilateral four-body problem, where three primary bodies move in a Lagrangian configuration, a well-known solution to the three-body problem, and a fourth body is considered infinitesimal. The primary bodies have equal masses, and we describe the dynamics of the infinitesimal particle and identify some periodic orbits. Building on this line of work, in collaboration with other co-authors, I have published several additional papers on this problem, now considering different mass values for the primaries. Key results include an analysis of the number of equilibria, their linear stability, and bifurcations. We also employed Birkhoff normal form theory and the Kolmogorov-Arnold-Moser (KAM) Theorem to explore the nonlinear stability of linearly stable equilibrium points [29, 30]. Additionally, we provided a description of transfer orbits between the lobes of the Hill regions surrounding the primaries for appropriate Jacobi constants (see [4]).

My contributions to the study of central 4- and 5-body configurations are extensive, so I shall highlight a few of my most significant publications. In [5], we characterized the convex and concave central configurations with an axis of symmetry for the 4-body problem, specifically under the condition that the masses satisfy $m_1 = m_2 \neq m_3 = m_4$. Considering that an *equilic quadrilateral* is a quadrilateral with one pair of opposite sides having the same length, having angles of inclination whose sum is $2\pi/3$, and a *Hjelmslev quadrilateral* is a quadrilateral with two right angles at opposite vertices. We characterize the central configurations of the 4-body problem whose four positive masses lie at the vertices of equilic quadrilaterals [6], and also we show that any 4-body central configuration forming a Hjelmslev quadrilateral must be a right kite configuration, [7].

Stacked central configurations are central configurations where a subset of the points' masses also form a central configuration. In [12], in collaboration with Cornelio and Cors, we studied a family of stacked central configurations in a planar a 5-body problem with masses m_1, m_2 and m_3 (ordered from left to right) lying on the same line and masses m_4 and m_5 placed symmetrically with respect to the line containing the first three bodies. In addition, we further assumed that $m_1 = m_2$ and that the three collinear bodies are symmetrical with respect to mass m_2 in the middle. We prove that when the collinear bodies form an Euler central configuration of the 3-body problem with $m_1 = m_3$, there exists a new family of stacked five-body central configuration.

The paper [9] studies the planar 5-body central configurations whose four bodies are located at the vertices of a rhombus. We prove the existence of central configurations when one diagonal is nearly equal to the sides of the rhombus and when the two diagonals are nearly equal. An *equilateral pentagon* is a polygon in the plane with five sides of equal length. In the paper [10], we classify the central configurations of the 5-body problem for which the five bodies lie at the vertices of an equilateral pentagon with an axis of symmetry. More recently, in collaboration with Cornelio and Cors [11], it was proved the existence of central configurations of the 5-body problem, where the five bodies are in a symmetric house-shaped configuration, That is, when the five bodies are located at the vertices of a convex rectangular pentagon.

References

1. Albouy, A., Kaloshin, V.: Finiteness of central configurations of five bodies in the plane. Ann. of Math. (2) **176**(1), 535–588 (2012)
2. Alvarez, M., Llibre, J.: Restricted three-body problems and the non-regularization of the elliptic collision restricted isosceles three-body problem. Extracta Math. **13**(1), 73–94 (1998)
3. Alvarez, M., Llibre, J.: Heteroclinic orbits and Bernoulli shift for the elliptic collision restricted three-body problem. Arch. Ration. Mech. Anal. **156**(4), 317–357 (2001)
4. Alvarez-Ramírez, M., Barrabés, E.: Transport orbits in an equilateral restricted four-body problem. Celestial Mech. Dynam. Astronom. **12**(2), 191–210 (2015)
5. Alvarez-Ramírez, M., Llibre, J.: The symmetric central configurations of the 4-body problem with masses $m_1 = m_2 \neq m_3 = m_4$. Appl. Math. Comput. **219**(11), 5996–6001 (2013)
6. Alvarez-Ramírez, M., Llibre, J.: Equilic quadrilateral central configurations. Commun. Nonlinear Sci. Numer. Simul. **78**, 104872, 7 pp. (2019)
7. Alvarez-Ramírez, M., Llibre, J.: Hjelmslev quadrilateral central configurations. Phys. Lett. A **383**(2–3), 103–109 (2019)
8. Alvarez-Ramírez, M., Vidal, C.: Dynamical aspects of an equilateral restricted four-body problem. Math. Probl. Eng. **2009**, Art. ID 181360, 23 pp. (2009)
9. Alvarez-Ramírez, Cornelio, J.L., Cors, J.M.: Central configurations in the five-body problem: rhombus plus one. Qual. Theory Dyn. Syst. **20**(2), Paper No. 51, 13 pp. (2021)
10. Alvarez-Ramírez, M., Gasull, A., Llibre, J.: On the equilateral pentagonal central configurations. Commun. Nonlinear Sci. Numer. Simul. **112**, Paper No. 106511, 12 pp. (2022)
11. Alvarez-Ramírez, M., Cornelio, J. L., Cors, J. M.: Convex symmetric rectangular pentagon central configurations. Commun. Nonlinear Sci. Numer. Simul. **138**, Paper No. 108250, 10 pp. (2024)
12. Cornelio, J.L., Alvarez-Ramírez, M., Cors, J.M.: A family of stacked central configurations in the planar five-body problem. Celestial Mech. Dynam. Astronom. **129**(3), 321–328 (2017)
13. Euler, L.: De moto rectilineo trium corporum se mutuo attahentium. Novi Comm. Acad. Sci. Imp. Petrop. **11**, 144–151 (1767)

14. Farquhar, R.W.: The control and use of libration-point satellites, PhD thesis, Dept. of Aeronautics and Astronautics, Stanford University, Stanford, CA, USA (1968)
15. Giuliatti Winter, S.M., Winter, O.C., Vieira Neto, E., Sfair, R.: Stable regions around Pluto. Mon. Not. R. Astron. Soc. **430**, 1892–1900 (2013)
16. Hampton, M., Moeckel, R.: Finiteness of relative equilibria of the four-body problem. Invent. Math. **163**(2), 289–312 (2006)
17. Lagrange, J.L.: Oeuvres de Lagrange. Gauthier-Villars, Paris **6**, 272–292 (1873)
18. Meyer, K.R., Offin, D.C.: Introduction to Hamiltonian Dynamical Systems and the N-body Problem. Applied Mathematical Sciences, 3rd edn., vol. 90. Springer, New York (2017)
19. Michael Jr., W.H.: Considerations of the motion of a small body in the vicinity of the stable libration points of the Earth-Moon system. In: NASA TR-R-160 (1963)
20. Moeckel, R.: Central configurations. In: Central Configurations, Periodic Orbits, and Hamiltonian Systems. Adv. Courses Math. CRM Barcelona, pp. 105–167. Birkhäuser/Springer, Basel (2015)
21. Moulton, F. R.: The straight line solutions of the problem of N-bodies. Bull. Amer. Math. Soc. **13**, 324 (1906)
22. Murray, C.D., Dermott, S.F.: Solar System Dynamics. Cambridge University Press, Cambridge (1999)
23. Newton, I.: Newton Manuscript in Keynes Collection. King's College, Cambridge, UK. MSS 130.6, Book 3; 130.5, Sheet 3
24. Poincaré, H.: Les méthodes nouvelles de la mécanique céleste I-III, Gauthier. Villars, Paris (1982)
25. Smale, S.: Topology and mechanics II. The planar n-body problem. Invent. Math. **11**, 45–64 (1970)
26. Smale, S.: Mathematical problems for the next century. In: Mathematics: Frontiers and Perspectives, pp. 271–294. American Mathematical Society, Providence (2000)
27. Szebehely, V.: Theory of Orbits. The Restricted Problem of Three Bodies. Academic Press (1967)
28. Wintner, A.: The Analytical Foundations of Celestial Mechanics. Princeton Mathematical Series, vol. 5. Princeton University Press, Princeton (1941)
29. Zepeda Ramírez, J.A., Alvarez-Ramírez, M.: Equilibrium points and their linear stability in the planar equilateral restricted four-body problem: a review and new results. Astrophys. Space Sci. **367**(8), Paper No. 77, 12 pp. (2022)
30. Zepeda Ramírez, J.A., Alvarez-Ramírez, M., García, A.: Nonlinear stability of equilibrium points in the planar equilateral restricted mass-unequal four-body problem. Int. J. Bifur. Chaos Appl. Sci. Eng. **31**(11), Paper No. 2130031, 15 pp. (2021)

The Cage and Diameter Problems for Bipartite Biregular Graphs

Gabriela Araujo-Pardo

Abstract In this summary, I expose my work on the Degree/Diameter and Cage Problems for bipartite biregular graphs. This review is the second part of another one published previously in [G. Araujo-Pardo, Contributions on the Cage and Diameter Problems. Advances in the Mathematical Sciences - AWM Research Symposium, Minneapolis, MN, June 2022. Springer Nature (2024)], where I provided a detailed analysis of my contributions in the original case of these problems, which are regular and biregular graphs in one of their early generalizations. As in the previous work, I finish this summary by giving a brief list of topics I have worked on in the last six years.

Keywords Diameter problem · Cage problem · Weighted cages · Bipartite biregular graphs

1 Introduction to the Cage and Diameter Problems

I start with a short review of these two related issues in extremal graphs: *The Cage Problem* and *The Degree/Diameter Problem.*

The Cage Problem involves constructing regular simple graphs with a specified degree and girth, and with the minimum number of vertices. On the other hand, the *Degree/Diameter Problem* focuses on finding regular graphs with a fixed diameter and the maximum number of vertices. Both problems are closely related.

These two problems are interconnected because when the girth equals twice the diameter plus one, the lower bound on the graph's order in the Cage Problem aligns with the upper bound in the Degree/Diameter Problem. We call this bound the *Moore Bound*. In other words, if we denote the girth as g and the diameter as d, and

---

G. Araujo-Pardo (✉)
Instituto de Matemáticas-Campus Juriquilla, Universidad Nacional Autónoma de México, Juriquilla, Qro., Mexico
e-mail: garaujo@im.unam.mx

P. E. Harris et al. (eds.), *Women in Mathematics in Latin America*,
Research Perspectives Ghent Analysis and PDE Center 14,
https://doi.org/10.1007/978-3-031-97857-9_2

$g = 2d+1$, then $n_0(k, g) = M_1(k, d)$, where $n_0(k, g)$ denotes the minimum number of vertices of a k-regular graph of girth g while $M_1(k, d)$ denotes the maximum number of vertices of a k-regular graph of diameter d. For more information about these two problems, the reader can consult the following surveys [19, 24].

In this summary, I give a review of my work on Bipartite Biregular Graphs, divided into two different sub-topics: the Moore (or diameter) problem [6, 7] and the Cage (or girth) problem [10, 12].

Finally, I mention the topics I have worked on recently and introduce the last generalization I have given, the concept of weighted cages [9].

It is important to emphasize that the figures that appear in this summary are also published in the cited papers.

2 Preliminaries

As in the first summary (see [1]), I will present the key tools essential for comprehending the techniques employed in the referenced papers.

Let $k \geq 2$ and $d \geq 1$, it is well known that the Moore bound for the Degree/Diameter problem of a k-regular graph of diameter d is equal to

$$M_1(k, d) = 1 + k + k(k-1) + \cdots + k(k-1)^{d-1} = \frac{k(k-1)^d - 2}{k-2}$$

It is obtained by counting a rooted tree's vertices in a vertex v of degree k and high d. Here, all the vertices have degree k except the leaves.

On the other hand, related to the girth problem, when $g = 2d + 1$ and d is the diameter, the same tree gives us a lower bound for a k-regular graph of girth g, denoted by $n_0(k, g) = n_0(k, 2d + 1) = M_1(k, d)$. Notice that here, the girth is odd.

When the girth g is even, the lower bound is calculated by counting the vertices that emerge of an edge uv and all of them are at distance $\frac{g-2}{2}$ of either u or v. Moreover, we can see this tree as two trees of high $\frac{g-2}{2}$ rotted one in u and the other in v (the vertices u and v have degree $k-1$, the rest of the vertices have degree k and the leaves degree one).

$$n_0(k, g) = \begin{cases} 1 + k + k(k-1) + \cdots + k(k-1)^{(g-3)/2} & \text{if } g \text{ is odd;} \\ 2(1 + (k-1) + \cdots + (k-1)^{g/2-1}) & \text{if } g \text{ is even.} \end{cases} \quad (1)$$

The people who work on cages called this bound *Moore Bound*, and the cages that attain this bound are called *Moore Cages*, but strictly the Moore Bound was initially given for the Degree/Diameter Problem. In this review, to avoid confusion, I will call them *Diameter Moore Bound* and *Cage Moore Bound* and denoted $M_1(k, d)$ and $n_0(k, d)$ respectively. Recall that $M_1(k, d) = n_0(k, 2d + 1)$.

In [23, 24], you can find more about the existence of Moore diameter/cages graphs, respectively. If we have a k-regular graph of diameter d and prove that it has maximum order, this order is denoted as $M(k, d)$ that is, the maximum order overall k-regular graphs of diameter d. Similarly, if we have a k-regular graph of girth g and prove that there is no other graph with fewer vertices, then G is a *cage* of order $n(k, g)$. We usually call the *defect* or the *excess* to the difference between $M(k, d)$ and $M_1(k, d)$, and between $n_0(k, g)$ and $n(k, g)$.

We start to give some preliminary tools and definitions that we will use along with this paper. First, we recall the relationship between the Moore cages of even girth and the finite geometries. It is known that these cages exist for any $k \geq 2$ and $g = 4$, they are the complete bipartite graphs $K_{k,k}$. Additionally, when $(k - 1)$ is a prime power and $g \in \{6, 8, 12\}$ these graphs are the incidence graphs of certain finite geometries or generalized polygons, specifically generalized $(g/2)$-gons of order $(k - 1)$ (see [20]). This is the main reason why (k, g)-cages with $n_0(k, g)$ vertices and even girth g are also referred to as *generalized polygon graphs* [17].

In particular, a 3-gon is also known as a *projective plane*. The 4-gons are called *generalized quadrangles*, and, the 6-gons are known as *generalized hexagons*. All these objects are known to exist for all prime power values of $(k - 1)$, and we do not have examples in another case (when $(k - 1)$ is not a prime power (cf. e.g. [25, p. 25])).

The incidence graph of a projective plane of order q for q prime power is a $(q + 1)$-regular graph of order $2(q^2 + q + 1)$; it is a $(q + 1, 6)$-Moore Cage.

2.1 Block Designs and Generalized Polygons with Two Parameters

In this subsection, I will introduce two combinatorial structures that will have an equivalent relationship with the first topic of this summary, the *Bipartite Biregular Cages:* bipartite biregular graphs with a minimum order, where each partite set has a different degree (introduced in [10]), as the relationship between the generalized polygons with the usual cages that I described before.

I will introduce the notion of Steiner systems and the definition and existence of generalized polygons with two parameters.

2.1.1 Steiner Systems

The definitions given here are also provided, for example, in [18], it is important to say that we also use the same reference in [10].

Let $t,\ v,\ k,\ \lambda$ be integers with $v \geq k \geq t \geq 2$ and $\lambda > 0$. A *t-(v, k, λ)-design* is a collection of k-subsets (*blocks*) of a v-set S such that every t-subset of S is contained in exactly λ blocks.

A Steiner system is a t-design with $\lambda = 1$, that is, a *Steiner system* $S(t, k, v)$ is a collection of k-subsets (blocks) of a v-set S such that every t-subset of S is contained in exactly one block. We shall use $S(t, k, v)$ as an equivalent to t-$(v, k, 1)$-design. A Steiner system $S(2, 3, v)$ is called a *Steiner triple system* and is denoted by $STS(v)$. The numbers v, k, λ, the number b of blocks, and the number r of blocks each element occurs in, satisfies the following well-known equalities:

$$r(k-1) = \lambda(v-1), \tag{2}$$

$$vr = bk. \tag{3}$$

Moreover, if a Steiner system exists, it is also necessary that:

$$v \equiv 1 \pmod{k(k-1)} \text{ or } v \equiv k \pmod{k(k-1)}. \tag{4}$$

It is well known that Steiner systems exist for certain specific values of k and v, the general question of for which values of k a Steiner system exists is an open problem; the congruences (4) provide necessary, but not sufficient, conditions. It is important to note that Kirkman proved that when $k = 3$ these conditions are both necessary and sufficient, that is a $STS(v)$ exists if and only if $v \geq 7$ is congruent to 1 or 3 modulo 6 [18].

2.1.2 Generalized Polygons

The last ingredient that we have to define to explain the work in this summary is the existence of "non-regular" generalized polygons and their conditions. On these generalized polygons, we have that in any point is incident to the same number of lines, and any line is incident to the same number of points, but these two parameters are different. The following definitions are given in [22]:

Definition 1 ([22]) Let $\mathcal{G} = (\mathcal{P}, \mathcal{L}, I)$ be finite point-line incidence geometry and $n \geq 2$ be an integer.

A chain of length h is a sequence of $h + 1$ elements $x_0, x_1 \ldots, x_h \in \mathcal{P} \cup \mathcal{L}$ such that

$$x_0 I x_1 I \ldots I x_h.$$

We said that the chain joins x_0 and x_h and that $\mathcal{S}$ is connected if for any two elements of $\mathcal{P} \cup \mathcal{L}$ there exists a chain joining them. The distance of the two elements x and y of $\mathcal{P} \cup \mathcal{L}$, denoted by $d(x, y)$, is the length of the shortest chain joining them.

Then, $\mathcal{G}$ is called a *generalized n-gon* if it is connected and satisfies the following axioms:

- $d(x, y) \leq n$ for all $x, y \in \mathcal{P} \cup \mathcal{L}$.
- If $d(x, y) = k < n$, then there exists a unique chain of length k joining x and y.
- For all $x \in \mathcal{P} \cup \mathcal{L}$ there exists $y \in \mathcal{P} \cup \mathcal{L}$ such that $d(x, y) = n$.

Definition 2 ([22]) The incidence graph (sometimes called *Levi Graph*) of a point-line incidence geometry is a bipartite graph whose two classes of vertices correspond to the set of points and the set of lines of the geometry, and two vertices are adjacent if and only if the corresponding point-line is incident in the geometry.

The following results are given in [26] and also we report these definitions and results in [10]:

Lemma 1 ([26]) *A geometry $\mathcal{G} = (\mathcal{P}, \mathcal{L}, \mathbf{I})$ is a generalized n-gon if and only if the incidence graph of $\mathcal{G}$ is a connected graph of diameter n and girth 2n in which each vertex is incident with at least three edges.*

Every generalized polygon $\mathcal{G}$ can be associated with a pair (s, t), called the *order* of $\mathcal{G}$, such that every line is incident with $s + 1$ points, and every point is incident with $t + 1$ lines [26]. This means, in particular, that the incidence graph of $\mathcal{G}$ is a bipartite biregular graph with the two degrees $s + 1$ and $t + 1$.

In 1964, Feit and Higman proved that finite generalized n-gons exist only for $n = \{3, 4, 6, 8\}$. When $n = 3$, we have the projective planes and in this case, there exists only for (q, q) with q a prime power; when $n = 4$, we have the generalized quadrangles, which are known to exist for parameter pairs $(q, q), (q, q^2), (q^2, q), (q^2, q^3), (q^3, q^2), (q - 1, q + 1), (q + 1, q - 1)$, where q is a prime power; when $n = 6$, we have the generalized hexagons with parameters $(q, q), (q, q^3), (q^3, q)$, where q is a prime power; and finally, for $n = 8$, we have the generalized octagons, which are only known to exist for the pairs $(q, q^2), (q^2, q)$, where q is an odd power of 2. In addition, the following result determines the orders of the two partite sets.

Theorem 1 ([26]) *If there exists a $\mathcal{G} = (\mathcal{P}, \mathcal{L}, \mathbf{I})$ generalized n-gon of order (s, t) for $n \in \{3, 4, 6, 8\}$ then:*

- *If $n = 3$, $|\mathcal{P}| = s^2 + s + 1$ and $|\mathcal{L}| = t^2 + t + 1$.*
- *If $n = 4$, $|\mathcal{P}| = (1 + s)(1 + st)$ and $|\mathcal{L}| = (1 + t)(1 + st)$.*
- *If $n = 6$, $|\mathcal{P}| = (1 + s)(1 + st + s^2t^2)$ and $|\mathcal{L}| = (1 + t)(1 + st + s^2t^2)$.*
- *If $n = 8$, $|\mathcal{P}| = (1 + s)(1 + st)(1 + s^2t^2)$ and $|\mathcal{L}| = (1 + t)(1 + st)(1 + s^2t^2)$.*

We will start with some specific biregular graphs. They are bipartite and biregular; any partite set should have a different degree.

3 Bipartite Biregular Graphs

In this section, we will study the Moore and Cage problems in some specific kind of biregular graphs: the bipartite biregular graphs.

Definition 3 A *Bipartite Biregular Graph* is a bipartite graph where each partite set has a different degree, and we call them *bbgraphs*.

I will present the results in the order in which these problems appeared in history. I will start by explaining my work with bipartite biregular Moore graphs and then with bipartite biregular cages. Furthermore, I will begin this work by stating that these two families of graphs coincide (and solve both problems) when they are the incidence graphs of Steiner systems (for diameter 3 and/or girth 6) or biregular generalized polygons (for diameter 4, 6, and 8 and/or girths 8, 12, and 16).

We will state the above more clearly next.

First, I introduce the bipartite biregular Moore graphs described by Fiol et al. in 1983 (see [27]), and second, the bipartite biregular cages introduced by Filipovski et al. in 2019 (see [21]).

Definition 4 ([27]) A *Bipartite Biregular Moore Graph* is a bipartite biregular graph of fixed diameter and maximum order. They are denoted by $[r, m; d] - bbmoore$ for $2 \leq r < m$ and diameter $d \geq 2$.

Definition 5 ([21]) A *Bipartite Biregular Cage* is a bipartite biregular graph of fixed girth and minimum order. They are denoted by $(r, m; g) - bbcages$ for $2 \leq r < m$ and even girth $g \geq 4$.

I enlist the two theorems that appear in the two papers cited in this work.

Theorem 2 ([10]) *Whenever a generalized quadrangle, hexagon, or octagon $\mathcal{G}$ of order (s, t) exists, its point-line incidence graph is an $(s+1, t+1; 8)$-, $(s+1, t+1; 12)$- or $(s+1, t+1; 16)$-bbcage, respectively.*

Hence, there exist, infinite families of bipartite biregular $(n+1, n^2+1; 8)$-, $(n^2+1, n^3+1; 8)$-, $(n, n+2; 8)$-, $(n+1, n^3+1; 12)$- and $(n+1, n^2+1; 16)$-bbcages.

The following theorem appears in [6] as a consequence of the previous theorem:

Theorem 3 ([6]) *There exist infinite families of bipartite biregular $[r^2+1, r+1; 4]$-, $[r^3+1, r^2+1; 4]$-, $[r+2, r; 4]$-, $[r^3+1, r+1; 6]$- and $[r^2+1, r+1; 8]$-bbmoore graphs.*

3.1 Bipartite Biregular Moore Graphs

In this subsection, I will explain some results that appear in the following papers [6, 7].

First of all, I will give a result related to the maximum order of a bipartite biregular graph with degrees r and s, and diameter d, known as a *Moore-like bound* and denoted by $M(r, s; d)$, given by Yebra et al. [27].

Theorem 4 ([27]) *Let $r \geq s \geq 3$, then $M(r, s; d)$ depending of the parity of d, then:*

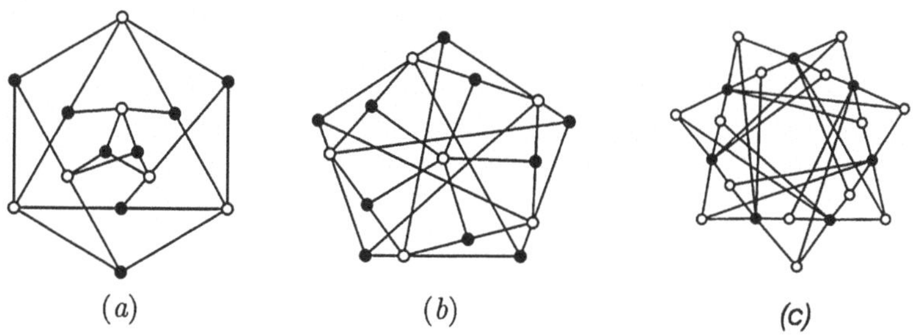

Fig. 1 (**a**) The [4, 3; 3]-bbmoore graph of order $M(4, 3; 3) = 14$. (**b**) The [5, 3, 3]-bbmoore graph of order $M(5, 3, 3) = 16$. (**c**) The [6, 3; 3]-bbmoore graph of order $M^*(6, 3; 3) = 21$

(i) If $d = 2m$ *and* $m \geq 1$ *then:*

$$M(r, s; 2m) = (r+s)\frac{[(r-1)(s-1)]^m - 1}{(r-1)(s-1) - 1}. \tag{5}$$

(ii) If $d = 2m+1$, $m \geq 1$, $\rho = \frac{r}{\gcd\{r,s\}}$ *and* $\sigma = \frac{s}{\gcd\{r,s\}}$, *then:*

$$M(r, s; 2m+1) = \left\lfloor \frac{1 + s(r-1)\frac{[(r-1)(s-1)]^m - 1}{(r-1)(s-1)-1}}{\rho} \right\rfloor (\rho + \sigma). \tag{6}$$

In Fig. 1a appears the [4, 3; 3]-bbmoore graph which has 6 vertices of degree 4 and 8 of degree 3 and in (b) the [5, 3; 3]-bbmoore graph which has 6 vertices of degree 5 and 10 of degree 3. Both were given in [27] and attain the Moore-like bound of Theorem 4.

Notices that for $d = 3$ and $r = 6$ and $s = 3$ the $M(6, 3; 3) = 24$, but in [6] we improve the previous bound for $d = 3$:

Proposition 1 ([6]) *If* $\rho = \frac{r}{\gcd\{r,s\}}$ *divides* $s - 1$, *then there is no* $[r, s; 3]$*-graph with order attaining the Moore-like bound in* (6).

Instead, the new improved Moore bound is equal to:

$$M^*(r, s; 3) = (1 + s(r-1) - \rho)\left(1 + \frac{\sigma}{\rho}\right). \tag{7}$$

Moreover, if G *is a* $[r, s; 3]$*-bbmoore that attains the new* $M^*(r, s; 3)$, V_1 *is the set of vertices of degree* r *and* V_2 *the set of vertices of degree* s *we have that:*

$$|V_1| \leq [1 + s(r-1) - \rho]\frac{\sigma}{\rho}, \qquad \text{and} \qquad |V_2| \leq 1 + s(r-1) - \rho.$$

As a consequence of Proposition 1 we proved that:

Corollary 1 ([6]) *For $s \geq 2$, the best Moore bounds for the orders $|V_1|$ and $|V_2|$ of a $[\rho s, s; 3]$-graph are as follows:*

(i) If $\rho | (s-1)$, then

$$|V_1| \leq (s^2-1) - \frac{s-1}{\rho}, \qquad \text{and} \qquad |V_2| \leq \rho(s^2-1) - (s-1).$$

(ii) If $\rho \nmid (s-1)$, then

$$|V_1| \leq s^2 - \lceil s/\rho \rceil, \qquad \text{and} \qquad |V_2| \leq \rho(s^2 - \lceil s/\rho \rceil).$$

In particular, we proved that a [6, 3; 3]-bbmoore graph has at most 21 vertices, with 7 vertices of degree 6 and 14 vertices of degree 3.

There exist different techniques to construct bbgraphs of diameter 3, 4, and 6; one of them consists of identifying some sets of the incidence graphs of the generalized polygons, for example, taking k incidence graphs of generalized polygons and identifying the vertices given by the lines, we obtain the following graphs, also given in [6]. Moreover, we also proved some results related to spectral graph theory.

Let G be a bipartite graph with stable sets V_1 and V_2 then, the *k-tuple (-V_1) graph* G^{kV_1} is obtained from G replacing each vertex of $u \in V_1$ of G by k vertices $u_1, \ldots, u_k$ with the same adjacences of u. Let V_1^k be the set of vertices with all the replacing vertices of V_1. The graph G^{kV_1} is bipartite with stable sets $V_1^k \cup V_2$ and satisfies the following result:

Theorem 5 ([6]) *Let $G = (V_1 \cup V_2, E)$ be a bipartite graph on $n = n_1 + n_2 = |V_1| + |V_2|$ vertices, diameter $D(\geq 2)$, and spectrum* $\operatorname{sp} G$. *Then, its k-tuple graph G^{kV_1}, on $N = kn_1 + n_2$ vertices, has the same diameter D, and spectrum*

$$\operatorname{sp} G^{kV_1} = \sqrt{k} \cdot \operatorname{sp} G \cup \{0^{(k-1)n_1}\}. \tag{8}$$

The following Theorem also appears in [6], but in that case, $k = 2$. I state it here for any integer $k \geq 2$; the proof is the same.

Theorem 6 *The following are $[r, kr; d]$-biregular bipartite graphs for $r \geq 3$, $r-1$ a prime power, and diameter $d \in \{3, 4, 6\}$.*

(i) A $[r, kr; 3]$-biregular bipartite graph has order $n = (k+1)(r^2 - r + 1)$.
(ii) A $[r, kr; 4]$-biregular bipartite graph has order $n = (k+1)(r^3 - 2r^2 + 3r)$.
(iii) A $[r, kr; 6]$-biregular bipartite graph has order $n = (k+1)(r^5 - 4r^4 + 7r^3 - 6r^2 + 3r)$.

In particular, applying Theorem 6 identifying two incidence graphs of a projective plane of order 2 (the (3; 3)-cage, also known as *Heawood Graph*), we obtain a [6, 3; 3]-bbmoore graph of order 21 with 7 vertices of degree 6 and 14 vertices of degree 3. This graph is given in Fig. 1c.

Notice that the bound $M^*(9, 3; 3) = 32$ is obtained applying 1. In the other case, using the Theorem 6 we obtain a [9, 3; 3]-bbgraph identifying the three incidence graphs of projective planes of order 2 ((3; 3)-cages or Heawood graphs) of order 28.

In a recently submitted paper [7], we improved, one step more, the Moore Bound and we can use this result to prove that the obtained [9, 3; 3]-bbgraph is also a cage.

Proposition 2 ([7]) *Given* $s \geq 3$ *and* ρ *such that* $s - 1 \leq \rho \leq s^2 - s - 3$, *there is no* $[\rho s, s; 3]$*-bigraph with order attaining the Moore bound in* (7), $M(\rho s, s; 3) = (s^2 - 1)(\rho + 1)$. *Instead, the new, improved Moore bound is*

$$M^*(\rho s, s; 3) = (s^2 - 2)(\rho + 1), \tag{9}$$

obtained with

$$|V_1| = s^2 - 2 \qquad \text{and} \qquad |V_2| = \rho(s^2 - 2). \tag{10}$$

In the rest of [7] we obtain similar results, with more general structures, called *difference sets*. It is well known that the existence of a projective plane of order a prime power $q = p^r$ depends on the existence of a different set, but we construct some bbgraphs with subsets of difference sets when the projective plane does not exist.

3.2 Bipartite Biregular Cages

In this subsection, I will explain some results that appear in the following papers [10, 12].

Recall that a bipartite biregular cage is a bipartite biregular graph of fixed girth and minimum order. And that they are denoted by $(r, m; g)$-bbcages for $2 \leq r < m$ and even girth $g \geq 4$.

As before, I will give a result related to the minimum order of a bipartite biregular graph with degrees m and n, and girth g, known as *Adjusted Moore lower bound* and denoted by $B(m, n; d)$, with $2 \leq m \leq n$, given by Filipovski et al. (see [21]).

Theorem 7 ([21]) *The* adjusted Moore lower bound $B(m, n; 2r)$ *for the orders of the bipartite biregular* $(m, n; 2r)$*-cages, depending of the parity of* r, *are the following:*

$$\begin{aligned} &1 + (n-1) + (n-1)(m-1) + \cdots + (n-1)^{\frac{r-1}{2}}(m-1)^{\frac{r-1}{2}} + \\ &1 + (m-1) + (m-1)(n-1) + \cdots + (m-1)^{\frac{r-1}{2}}(n-1)^{\frac{r-1}{2}}, \end{aligned} \tag{11}$$

when r is odd, and

$$\begin{aligned} &1+(n-1)+(n-1)(m-1)+\cdots+(n-1)^{\frac{r}{2}}(m-1)^{\frac{r}{2}-1}+ \\ &1+(m-1)+(m-1)(n-1)+\cdots+(m-1)^{\frac{r}{2}}(n-1)^{\frac{r}{2}-1}, \end{aligned} \tag{12}$$

when r is even.

Also in [21] the authors proved that, if $r = 2$, the bound (12) is sharp for all pairs m, n, as the smallest bipartite biregular $(m, n; 4)$-graphs are the complete bipartite graphs $K_{m,n}$. Also, defined, as usual, the difference between the order of a bipartite biregular $(m, n; 2r)$-graph Γ and the corresponding adjusted Moore bound $B(m, n; 2r)$ as the *excess* of Γ.

In [10] we improve the previous bound for odd r:

Theorem 8 ([10]) *Let Γ be a bipartite biregular $(m, n; 2r)$-graph, and suppose that $r \geq 3$ is odd, $m < n$, then $B(m, n)$ is equal to:*

$$1+n+n(m-1)+n(m-1)(n-1)+n(m-1)^2(n-1)+\cdots+n(m-1)^{\frac{r-1}{2}}(n-1)^{\frac{r-3}{2}}+$$

$$\begin{cases} \dfrac{n(m-1)^{\frac{r-1}{2}}(n-1)^{\frac{r-1}{2}}}{m} & or \\ \left\lceil \dfrac{n(m-1)^{\frac{r-1}{2}}(n-1)^{\frac{r-1}{2}}}{m} \right\rceil + \left\lceil \frac{n}{m} \right\rceil . \end{cases}$$

Depending if $\frac{n(m-1)^{\frac{r-1}{2}}(n-1)^{\frac{r-1}{2}}}{m}$ is an integer number or not.

In particular, we explore the problem for $g = 6$ and prove the following Theorems 9 and 10: Together, they describe a result as significant as the classical result, which states that the Moore r-regular cages of girth 6 exist if and only if the projective planes of order $q - 1$ exist.

In [10], we prove that the bipartite biregular cages of girth 6 that achieve the bound given in Theorem 8 exist, if and only if, the Steiner systems exist.

Theorem 9 ([10]) *The point-block incidence graph Γ of a Steiner system $S(2, k, v)$, $k \geq 3$, is a bipartite biregular $(k, \frac{v-1}{k-1}; 6)$-cage.*

Theorem 10 ([10]) *If Γ is a bipartite biregular $(k, n; 6)$-cage of order $\frac{(n+k)(nk-n+1)}{k}$, the pair $(S_\Gamma, \mathcal{B}_\Gamma)$ forms a Steiner system $S(2, k, 1 + n(k-1))$.*

In this paper, we obtain other important results for bipartite biregular cages of girth 6 obtained deleted points and lines in the incidence graphs of the projective planes, in particular, we constructed an $(6, 7; 6)$-bbgraph of order 91 and we prove that it is a $(6, 7; 6)$-bbcage using some divisibility criterion when the projective plane does not exists.

Finally, also in that paper, we solved the problem for $m = 3$ and $g = 6$ using the Steiner triple systems. In particular, we proved that:

Theorem 11 ([10]) *Let $n \geq 3$ be an integer.*

1. *If* $n \equiv 0, 1 \pmod 3$, *then a* $(3, n; 6)$*-bipartite biregular cage is the point-block incidence graph of the Steiner triple system* $STS(2n+1)$, *and its order is* $1 + \frac{n(2n+7)}{3}$.
2. *If* $n \equiv 2 \pmod 3$, *then the order of a* $(3, n; 6)$*-bipartite biregular cage is* $2 + \frac{n(2n+8)}{3}$.

Moreover, in that paper, we also proved that the construction given in Theorem 11, gives us a general construction of a $(m, n, 6)$-bipartite biregular graph, for $n \equiv -1 \pmod m$:

Theorem 12 ([10]) *If* $n \geq 4$ *and an* $S(2, n, v)$ *exists, the order of an* $(n, \frac{v-1}{n-1} - 1; 6)$*-bipartite biregular cage is bounded from above by:*

$$v + \frac{v(v-1)}{n(n-1)} - \frac{v-1}{n-1} - 1.$$

And, in [12], we proved, using arithmetic arguments, that this graph is generally a $(m, n, 6)$-bipartite biregular cage.

Theorem 13 ([12]) *Let* $3 \leq m \leq n$ *and* $n \equiv -1 \pmod m$. *Then the size of the* $(m, n; 6)$*-bipartite biregular cage is at least* $(\frac{n}{m} + 1)(n+1)(m-1)$ *and if a Steiner system* $S(2, k = m, v = 1 + n(m-1) + m)$ *exists, one can obtain a cage of this size by deleting a point and all the blocks through the deleted point.*

Moreover, also in [12] we obtain different $(m, n; 2r)$-graphs for $g = \{8, 12\}$ using the existence of generalized r-gons and deleting some subsets of points and lines on them:

Theorem 14 ([12]) *Suppose that there exists a generalized* r*-gon of order* (m, n). *Then there exist bipartite biregular graphs with parameters* $(m, n+1; 2r)$ *and* $(m+1, n; 2r)$ *of order*

$$n^{\frac{r}{2}-1} m^{\frac{r}{2}-1} (m + n + 1).$$

In particular, if $m = n$ *we obtain graphs of order* $n^{r-2}(2n+1)$.

In [12], we give a table with the parameters of the known finite generalized $2r$-gons of order (m, n) with $1 < m \leq n$ and $r \geq 4$, the order of the graphs constructed in Theorem 14, the corresponding Moore adjusted bound, and the excess of the graph.

Finally, in that paper, we can improve the results for girth $g = 8$, taking into account specific substructures on the generalized polygons:

Theorem 15 ([12]) *Let* $4 \leq m \leq n$ *be integers and suppose that there exists a generalized quadrangle* $\mathcal{G}$ *of order* (m, n). *Let* $m_1 \leq n_1$ *be integers satisfying the inequalities* $m_1 \leq m$ *and* $n_1 \leq n$. *Then there exist* $(m_1, n_1; 8)$*-bbgraphs of order* $(m_1 + n_1)mn$.

Finally, one of the contributions of this article is to take into account point or line structures in generalized quadrangles, called *ovoids* (maximum sets of points or lines in which no two points are collinear) or *spreads* (maximum sets of lines with no incident lines) and *subquadrangles* that are subsets in the generalized quadrangles that preserve inside of them the properties of generalized quadrangles.

Theorem 16 ([12]) *Suppose that there exists a generalized quadrangle* $\mathcal{G}$ *of order* (m, n) *which contains a subquadrangle* $\mathcal{G}'$ *of order* $(m, n/m)$. *Then there exists an* $(m, n+1; 8)$*-bbgraph of order* $(m+n+1)\frac{m^2-1}{m}n$.

It is important to say that these subquadrangles exist only for some specific parameters.

Using the following important result given in [10] that states a new lower bound $B_c(m, n; 2r)$ when the generalized r-gon does not exist.

Corollary 2 ([10]) *Let* $3 \le m \le n$, $r \ge 4$, *and suppose that no generalized* r*-gon of order* $(m-1, n-1)$ *exists. Then*

$$B_c(m, n; 2r) \ge B(m, n; 2r) + \frac{m+n}{\gcd(m, n)}.$$

And, regarding the non-existence of generalized polygons of some parameters, we give some lower bounds about the existence of bipartite biregular graphs:

Theorem 17 ([12]) *Let* $3 \le m$ *be an integer. Then*

(i) $B_c(m, m+1; 8) \ge (2m+1)(m^2-m+2)$;
(ii) $B_c(m, m^2+1; 8) \ge (m^2+m+1)(m^3-m^2+2)$;
(iii) $B_c(m, m+1; 12) \ge (2m+1)(m^4-2m^3+2m^2-m+2)$.

Using these previous bounds and interesting geometric properties given also in [22] related to elliptic, parabolic, and hyperbolic quadrics on some specific generalized quadrangles, the paper states the following results:

Theorem 18 ([12]) *If* q *is a prime power, then there exist a* $(q, q+1; 8)$*-bbgraph of order* $(2q+1)(q^2-1)$.

In particular, for $q = 3$ *there exists a* $(3, 4; 8)$*-bbcage on* 56 *vertices.*

Theorem 19 ([12]) *If* q *is a prime power, then there exist a* $(q, q^2+1; 8)$*-bbgraph of order* $(q^2+q+1)(q^3-q)$.

Theorem 20 ([12]) *If* q *is a prime power, then there exist a* $(q, q+1; 12)$*-bbgraph of order* $(2q+1)(q^4-q)$.

4 More Work on Generalizations of Cages

As I detailed in [1], in the last years I have worked on more topics related to the Degree/Diameter and Cage problems: *mixed Moore graphs* [4], *mixed cages* [2, 5, 8], *planar regular and biregular cages* [11], *edge-girth-regular graphs* and *almost-girth-regular graphs* [3, 14], *(r, g, χ)-graphs* [13, 16] and *girth-diameter graphs* [15].

Finally, I would like to include a brief description of the latest work I have done as a generalization of the cage problem, and these are the *weighted cages*. I introduce this concept in [9] in collaboration with C. De la Cruz, M. Matamala and M. Pizaña.

A *weighted graph*, introduced in [9], is a (simple, finite) graph $G = (V, E)$ together with a weight function $w : E(G) \to \mathbb{R}$, however, to keep the presentation as simple as possible, we shall focus on weight functions of the form $w : E(G) \to \{1, 2\}$. An edge with weight 1 is called a *light edge* while one with weight 2 is called a *heavy edge*.

The paper has recently been submitted for review, and in the paper, we give a pair of motivations to define these new cages. Just as we generalized the cage problem to directed or mixed graphs, it is natural to extend it to weighted graphs. One of the areas where this extension was natural is in traversability, in Eulerian and Hamiltonian graphs, as well as in the well-known problems of the Chinese Postman and the Traveling Salesman. Moreover, we can also say that this generalization "distinguishes" between geometry and combinatorics. For example, as we will see later, the mixed graph can have a triangle, yet the girth may vary between 3 and 6 depending on the weights of the edges of the triangle.

Another motivation is directly related to the relationship between the two problems that I explain in this summary: the *Cage Problem* and the *(Δ, d)-problem*, it then becomes natural to think about Moore graphs with different weights as edges with higher weights could imply, depending on the problem, "shortcuts" or "delays" between two vertices when trying to reach all vertices with a fixed distance.

As [9] I only want to share that we define the notion of weighted cages, prove their existence, and give some *Moore like-bounds*, using a similar but more sophisticated tree structure that we used to give the usual lower bound for regular cages, characterizes the weighted cages of girth 3 and 4, and has computerized, exhaustive searches using backtracking with symmetry reductions to obtain values of $n(a, b, g)$. Finally, we give general bounds using the incidence graphs (or cages) associated with finite geometries.

Acknowledgments As in [1], I would like to express my heartfelt gratitude to all the coauthors with whom I have written the papers whose results are described in this summary, thank you for sharing your mathematics with me to create results together. Once again, I emphasize the importance of collaborative and diverse work in research and across all aspects of mathematics.

Truly, I have learned a great deal from all of you—teachers, peers, students, and all of my friends.

Once again, I also have to mention that almost all the figures appearing in this summary were collected from the articles cited within it.

Finally, I would like to thank the comments and corrections of the referees, which notably improved this article.

Research supported by PAPIIT-UNAM-México IN113324 and CONAHCyT CBF2023-2020-552.

References

1. Araujo-Pardo, G.: Contributions on the cage and diameter problems and beyond. In: Advances in the Mathematical Sciences - AWM Research Symposium, Minneapolis, MN, June 2022. Springer Nature, Cham (2024). https://link.springer.com/book/9783031936685
2. Araujo-Pardo, G., Mendoza, L.: On mixed cages of girth 6 (submitted)
3. Araujo-Pardo, G., Leemans, D.: Edge-girth-regular graphs arising from biaffine planes and Suzuki groups. Discrete Math. **345**(10), Paper no. 112991 (2022). https://doi.org/10.1016/j.disc.2022.112991
4. Araujo-Pardo, G., Balbuena, C., Miller, M., Ždímalová, M.: A family of mixed graphs with large order and diameter 2. Discrete Appl. Math. **218**, 57–63 (2017)
5. Araujo-Pardo, G., Hernández-Cruz, C., Montellano-Ballesteros, J.J.: Mixed cages. Graphs Combin. **35**(5), 989–999 (2019)
6. Araujo-Pardo, G., Dalfó, C., Fiol, M.A., Lôpez, N.: Birregular bipartite Moore graphs. Discrete Math. **344**(11), Paper No. 112582, 12 pp. (2021). https://doi.org/10.1016/j.disc.2021.112582
7. Araujo-Pardo, G., Dalfó, C., Fiol, M.A., López, N.: Birregular Bipartite Moore Graphs. Discrete Math. 344, no. 11, Paper No. 112582, 12 pp. (2021). https://doi.org/10.1016/j.disc.2021.112582
8. Araujo-Pardo, G., De la Cruz, C., González-Moreno, D.: Mixed cages: Monotonicity, connectivity and upper bounds. Discrete Math. **345**(5), 112792 (2022)
9. Araujo-Pardo, G., De la Cruz, C., Matamala, M., Pizaña, M.: Weighted cages (In review). arxiv.org/abs/2411.02705
10. Araujo-Pardo, G., Ramos-Rivera, A., Jajcay, R., Szönyi, T.: On a relation between bipartite biregular cages, block designs, and generalized polygons. J. Combin. Des. **30**(7), 479–496 (2022). https://doi.org/10.1002/jcd.21836
11. Araujo-Pardo, G., Barrera-Cruz, F., García-Colín, N.: Biregular (and regular) planar cages. Discuss. Math. Graph Theory **43**(4), 1171–1194 (2023). https://doi.org/10.7151/dmgt.2425
12. Araujo-Pardo, G., Kiss, G., Szönyi, T.: New results on bipartite biregular cages, block designs, and generalized polygons. Bol. Soc. Mat. Mex. **31**, 20 (2025). arXiv.2305.03290v1. https://doi.org/10.1007/s40590-024-00700-w
13. Araujo-Pardo, G., Berikkyzy, Z., Lesniak, L.: On the existence of (r, g, χ)-Cages, Graphs and Combinatorics, (2025). https://doi.org/10.1007/s00373-025-02915-8
14. Araujo-Pardo, G., Kiss, G., Porupsánszki, I.: On extremal (almost) edge-girth-regular graphs. Art Discrete Appl. Math. (2025). https://doi.org/10.26493/2590-9770.1733.dc9. https://adam-journal.eu/index.php/ADAM/article/view/1733
15. Araujo-Pardo, G., Conder, M., García-Colin, N., Kiss, G., Leemans, D.: A note of girth-diameter cages. Art Discrete Appl. Math. (2025). https://doi.org/10.26493/2590-9770.1743.19f. https://adam-journal.eu/index.php/ADAM/article/view/1743
16. Araujo-Pardo, G., Díaz-Calderón, J.C., Fresán, J., González-Moreno, D., Lesniak, L., Olsen, M.: On (r, g, χ)-graphs and cages for infinitely many values of regularity r, girth g and chromatic number χ. Art Discrete Appl. Math. (2025). https://doi.org/10.26493/2590-9770.1737.0f8. https://adam-journal.eu/index.php/ADAM/article/view/1737
17. Biggs, N.: Algebraic Graph Theory. Cambridge University Press, New York (1996)
18. Colbourn, C.J., Dinits J. (eds.): The CRC Handbook of Combinatorial Designs. Discrete Mathematics and its Applications, 2nd edn., 984 p. Chapman & Hall/CRC, Boca Raton (2007)

19. Exoo, G., Jajcay, R.: Dynamic cage survey. Electron. J. Combin. Dynamic Survey **16**, 1–55 (2013)
20. Feit, W., Higman, G.: The non-existence of certain generalized polygons. J. Algebra **1**, 114–131 (1964)
21. Filipovski, S., Ramos-Rivera, A., Jajcay, R.: On biregular bipartite graphs of small excess. Discrete Math. **342**, 2066–2076 (2019)
22. Kiss, G., Szönyi, T.: Finite Geometries. CRC Press/Taylor & Francis Group, Boca Raton (2020)
23. Jajcay, R., Kiss, G., Miklavič, S.: Edge-girth-regular graphs. Eur. J. Combin. **72**, 70–82 (2018)
24. Miller, M., Širáň, J.: Moore graphs and beyond: a survey. Electron. J. Combin. Dyn. Surv. **20**(2) (2013), #DS14v21
25. Payne, S.E., Thas, J.A.: Finite Generalized Quadrangles. Research Notes in Mathematics, vol. 110. Pitman (Advanced Publishing Program), Boston (1984)
26. H. Van Maldeghem Generalized Polygons. Birkhäuser, Springer, Basel (1998)
27. Yebra, J.L.A., Fiol, M.A., Fàbrega, J.: Semiregular bipartite Moore graphs. Ars Combin. **16A**, 131–139 (1983)

A Measure of Financial Technologies Usage: A Proposal for Mexico

Verónica Arredondo, Brenda Valeria López Falcón, and Leticia Adriana Ramírez Hernández

Abstract Technology has undoubtedly improved the efficiency of daily activities and revolutionized human interaction in social life. This research focuses on analyzing advances in financial technology (fintech) usage and proposes an index to measure its adoption in Mexico. The proposed indicator includes two key variables from the National Survey on Availability and Use of Information Technologies in Households (ENDUTIH) 2022, conducted by the National Institute of Statistics and Geography (INEGI), (Marco geo-estadístico. Technical report, INEGI & IFT, 2024). The new indicator evaluates fintech usage across Mexican states, including Mexico City, referred to as federal entities.

Keywords Index · Fintech · Territory · Social · Mexico

1 Introduction

The State as a social regulatory entity is required to facilitate the well-being of individuals and society. In accordance, the State establishes regulations, rules, and policies that, in theory, seek the improvement and comprehensive development of the individuals who make up a community. Considering minority groups [1], as well as differences in socioeconomic levels and their effects across various fields, is fundamental to collective progress.

Nowadays, technological advances and computer tools are essential instruments in daily life. These tools are required to process and resolve the problems that

V. Arredondo (✉) · L. A. Ramírez Hernández
Unidad Académica de Matemáticas - Universidad Autónoma de Zacatecas, Zacatecas, Zac., Mexico
e-mail: varredondo@uaz.edu.mx; lramirez@uaz.edu.mx

B. V. López Falcón
Instituto Nacional de Geografía e Historia (INEGI), Aguascalientes, Aguascalientes, Mexico
e-mail: Valeria.lfalcon@gmail.com

P. E. Harris et al. (eds.), *Women in Mathematics in Latin America*,
Research Perspectives Ghent Analysis and PDE Center 14,
https://doi.org/10.1007/978-3-031-97857-9_3

citizens face regularly. Mobile phone applications help individuals find places to eat, access transportation (including ride-sharing platforms), and locate accommodations. They also provide information about the weather, event demand, and traffic congestion. Other gadgets have specific functions such as step counters, distance tracking, and calorie monitoring.

The financial sector is not lagging in adopting technological innovations. The term *fintech* refers to activities involving technological developments related to the design, offering, and presentation of financial products and services, [9]. The word "fintech" originates from the combination of "finance" and "technology." This concept began to be used when the first online banking and payment systems were launched in the 1990s. After the digitization of banking systems, *fintech 2.0* (the second stage following *fintech 1.0*) integrated mobile internet. *Fintech 3.0* introduced products offering broader solutions to users and customers, significantly enhancing their financial experience, [22].

Research findings indicate that in 2024 there are over 700 fintech initiatives in Mexico. This highlights the need to build a new financial culture that includes understanding these services, their uses, qualities, advantages, and disadvantages. This effort is essential to prevent the gap between early and late adopters of fintech from widening further. Efforts should be made to incorporate more users into the ecosystem [7].

The adoption of fintech offers new growth opportunities for countries and their citizens. Artificial intelligence, data, and statistics can help users make faster and more effective decisions based on their needs. The fintech narrative is defined by evolution, change, transformation, and adaptation. Thus, fintech redefines how we understand and use money [4].

In this context, the importance of measuring fintech usage has garnered considerable attention from researchers, who highlight the critical role that metrics and indexes play in understanding financial inclusion and empowerment.

Several studies emphasize the importance of financial services usage for the economic success of SMEs [13]. Demirgüç-Kunt et al. [6] highlight fintech's pivotal role in expanding access for underserved populations, advocating for comprehensive indexes to track usage and identify trends. Chen et al. [5] stress the transformative potential of digital financial services, noting the need for precise metrics to assess use and impact considering dimensions like availability, affordability, and usability. Carbo-Valverde et al. [3] call for robust measurement frameworks within traditional banking systems to evaluate fintech's impact and ensure benefits reach marginalized groups.

Collectively, these studies underscore the importance of developing and utilizing effective metrics and indexes to measure fintech usage and its impact on financial inclusion. Such measurements are essential not only for enhancing our understanding of financial inclusion but also for informing policy decisions, driving innovation, and ultimately empowering individuals and businesses to achieve better financial outcomes in an increasingly digital world. The consistent call for robust and context-specific measurement frameworks reflects the complexity of assessing the true impact of fintech and the need for ongoing research and data collection.

This study focuses on presenting a new proposal to measure fintech usage in Mexico, referred to as the Fintech Usage Index for Mexico. Additionally, it suggests further research involving statistical models to provide more information on how socioeconomic profiles influence fintech usage in Mexico. Extensions to this work could include considering the X-model, which generates synthetic variables that emulate original ones, allowing the separation of real effects from spurious relationships in the variable of interest, this study is described in detail in Vásquez et al., [21]. Moreover, future research could analyze behavioral trends over time, since this index can be estimated for different years using available ENDUTIH data.

It is worth it to mention that, the proposed Index is a measure suitable for assessing the use of fintech in the financial services sector, rather than in financial products. Financial services refer to money management and transactions, such as online banking, payments, financial advisory services, remittances, which facilitate transactions and asset management. In contrast, financial products include specific offerings like personal loans, credit cards, mutual funds, certificates of deposit (CDs), and various insurance policies, [15].

This index is more appropriate for evaluating fintech usage for financial services and not for banking products, because it focuses on how consumers engage with digital platforms that provide services such as payments, banking, and asset management.

This article is organized as follows: First, the introduction is presented. Section 2 describes the materials and data used in the research. Section 3 introduces the new proposal for an indicator of fintech usage in Mexico, including the mathematical formulation of the indicator and an explanation of its components. Section 4 presents the results of implementing the proposed index in Mexico and demonstrates its application on a map of Mexico. Lastly, conclusions, reflections, and potential extensions of the research are discussed.

2 Materials and Data

This section presents the tools used for data analysis. Additionally, it introduces the variables and defines the database utilized in this study.

The data considered in this research were sourced from the National Survey on Availability and Use of Information Technologies in Households ENDUTIH 2022, [11], a database generated by INEGI [12]. This database aims to analyze the use of information and financial technologies in Mexico.

As established by INEGI, the ENDUTIH survey seeks to:

- Obtain information on the availability and use of information and communication technologies in households and their usage by individuals aged six years and older in Mexico.
- Generate statistical information on these subjects to support decision-making regarding public policies.

- Provide elements of analysis for national and international studies as well as for the general public interested in the topic.

Regarding the inclusion criteria for this study and the records from the ENDUTIH survey related to the variables of interest were considered. The sampling method employed in the ENDUTIH survey was based on a probabilistic, stratified, and cluster sampling design. The Master Sample was used, a national housing framework constructed from data gathered in the 2010 Population and Housing Census. The primary sampling units (PSUs) were formed by groupings of housing with differentiated characteristics depending on the geographical area. This section presents the various tools that were used for data analysis. Correspondingly, the variables and the definition of the database used are presented. The data considered in this research were taken from the National Survey on Availability and Use of Information Technologies in Homes (ENDUTIH) 2022, a database generated by INEGI, to analyze the access and use of information technologies and financial technologies in Mexico, [11].

Considering the data available from the ENDUTIH 2022 survey, this study specifically focuses on records of individuals aged 18 or older. This limitation ensures the analysis is confined to individuals with autonomous usage of banking services.

For the proposed measure of financial technology (fintech) usage in Mexico, the following variables were selected:

- *Making payments online in the last 12 months:* This variable indicates whether respondents made online payments, either for purchases or services, during the last 12 months.
- *Installation of mobile banking applications:* This variable identifies whether respondents have installed mobile banking applications, such as those provided by Bancomer or Banamex (Officially Bancomer is named BBVA México, S.A., Institución de Banca Múltiple, and Banamex is officially named Grupo Financiero Citibanamex, S.A. de C.V.), on their mobile devices. The survey ENDUTIH 2022 explicitly mentions Bancomer or Banamex, [11].

The proposed index for measuring fintech usage in Mexico assigns greater weight to whether users have banking applications. This is because such apps represent a comprehensive tool that facilitates not only online payments but also a broader range of financial services, such as account management, transfers, and access to credit or investments. Having a banking app could demonstrate that users are equipped to interact with various fintech products and services, promoting greater financial inclusion. Furthermore, the indicator focuses on the supply and demand for banking services and emphasizing real-time transactions and money management. These services provide access to immediate payment solutions and financial tools as opposed to strategic banking decisions such as credit applications, which require deeper analysis and do not necessarily involve frequent interactions. The analysis and estimation of the proposed indicator and related statistical tests were conducted using the Statistical Package for the Social Sciences [10, 18, 19], and the Visual Studio Code development environment [14].

3 Proposal for an Indicator of Financial Technologies Usage in Mexico

The construction of the indicator is based on variables retrieved from the ENDUTIH 2022 database. The selected variables, such as online payments and the use of mobile banking applications, are weighted to introduce a measure of fintech usage. The mathematical formulation of the proposed indicator is presented in Eq. (1):

$$I_{FT} = \left(\sum_{i=1}^{n} \frac{x_{1i}}{P} W_1 + \sum_{i=1}^{n} \frac{x_{2i}}{P} W_2 \right) \times 10 \tag{1}$$

In which:

- I_{FT}: Fintech use index (financial technologies).
- x_{1i}: Value of Component 1 for i-th observation, representing the variable "making payments online in the last 12 months," weighted by the ENDUTIH 2022 expansion factor.
- x_{2i}: Value of Component 2 for i-th observation, representing the installation of mobile applications for accessing mobile banking, weighted by the ENDUTIH 2022 expansion factor.
- n: Number of observations within the considered region.
- P: Population of the territorial considered region ($\geq$18 years).

The definition weight values are:

- $W_1 = 0.40$: Weight of component 1.
- $W_2 = 0.60$: Weight of component 2.

The decision to weight the fintech usage indicator with 40% for online payments and 60% for mobile banking app installation reflects the greater importance of active engagement with digital financial services. Installing a mobile banking app (60%) demonstrates a deeper commitment, as it involves regularly using the app to manage finances, access a broader range of services, and engage more fully with financial technology. In contrast, making online payments (40%) is significant but might not indicate ongoing use or a strong connection to the financial services ecosystem. Therefore, prioritizing mobile app installation provides a more accurate representation of true fintech usage and user involvement in digital finance.

Certainly, the variables can be weighted in various ways; however, based on the available data, this specific weighting was deemed appropriate. This approach considers the significance of each variable in relation to the overall analysis and ensures that the assessment accurately reflects their contributions. By using this determined weighting, the goal is to create a reliable framework that captures the necessary insights while aligning with the empirical evidence at hand.

The proposed index to measure fintech usage in Mexico assigns scores of 0, 4, 6, or 10 for each individual, depending on their reported usage. The territorial indicator ranges from 0 to 10 points and represents the average usage of fintech within each territorial entity. A score of 0 indicates the lowest level of fintech usage, while a score of 10 reflects the highest level of engagement with digital financial services.

3.1 Proposed Model to Measure a Relationship Between Usage of Financial Technologies in Mexico and Sociodemographic Characteristics

This section presents a specific tool designed to study and analyze the relationship between the use of financial technologies (FinTech) in Mexico and sociodemographic characteristics. It incorporates a composite indicator (I_{FT}) to measure fintech usage, which is calculated using weighted variables from the ENDUTIH 2022 database, specifically internet usage and mobile banking application usage, [11]. The I_{FT} ranges from 1 to 4, representing low to high levels of fintech usage.

Furthermore, this section outlines a logistic proportional partial ordinal model—an extension of the logistic ordinal model—to explore the relationship between I_{FT} and variables such as sex, geographic location, education level, socioeconomic status, and age. The study aims to understand how these factors influence the use of financial technology within the Mexican context.

The proposed index to measure fintech usage in Mexico can take values of 0 (level 1), 4 (level 2), 6 (level 3), and 10 (level 4) for each individual. At the territorial level, the indicator ranges from 0 to 10 points, representing the average score per territorial entity. A score of 1 reflects the lowest level of fintech usage, while a score of 4 indicates the highest level of engagement with digital financial services.

To analyze the possible relationships between fintech usage levels and the sociodemographic profile, the following logistic proportional partial ordinal model is proposed, [16]:

$$I_{FT} = \beta_1 x_1 + \beta_2 x_2 + \beta_3 x_3 + \beta_{41} x_{41} + \beta_{42} x_{42} + \beta_{43} x_{43} + \beta_{44} x_{44} + \beta_5 x_5 + \varepsilon \quad (2)$$

Where:

- I_{FT}: Fintech usage index (financial technologies).
- x_1: Sex—A binary or categorical variable representing gender.
- x_2: Geographic zone (rural or urban)—a binary or categorical variable representing geographic location.
- x_3: Years of education—a continuous variable.
- x_{41}: Low socioeconomic level—a categorical variable indicating low socioeconomic status.
- x_{42}: Medium-low socioeconomic level—a categorical variable indicating medium-low socioeconomic status.
- x_{43}: Medium-high socioeconomic level—a categorical variable indicating medium-high socioeconomic status.

- x_{44}: High socioeconomic level—a categorical variable indicating high socioeconomic status.
- x_5: Age—a continuous variable.
- ϵ: Error term.
- $\beta_1, \beta_2, \beta_3, \beta_{41}, \beta_{42}, \beta_{43}, \beta_5$: Regression coefficients to be estimated.

In the present study, a regression model without an intercept term was chosen when incorporating the categorical variable 'socioeconomic level,' which consists of four categories. This decision allows for the direct estimation of the effect of each category on the dependent variable, avoiding redundancies and facilitating a clear interpretation of the results. According to the literature, excluding the intercept is appropriate when each category is intended to have its own coefficient, reflecting its specific impact [17, 20].

The logistic proportional partial ordinal model described is not directly represented by a single equation, as in linear regression. Instead, it employs a latent variable approach and cumulative probabilities [8]. While the document provides a linear predictor for convenience, this represents only part of the model. The probabilities are calculated as follows: The model estimation involves determining the most likely values for the β coefficients and the I_{FT} thresholds that best fit the observed data. The linear equation (2) shown is a simplification representing the linear predictor for the latent variable. The logistic proportional partial ordinal model for ordinal outcomes of I_{FT} accounts for the ranking of categories and estimates probabilities based on predictor variables.

To ensure valid statistical inference, the model considers independent observations, absence of multicollinearity, and partial proportional odds. These assumptions enable stable estimation and provide a clear interpretation of the model's results.

The estimation results and discussion for the proposed model will be presented as part of a further research. In fact, these results and their interpretations are detailed in the preprint Arredondo et al. [2].

4 Implementation and Results of the New Indicator Proposals

The results presented in Table 1 reveal the varying levels of FinTech adoption across Mexican states. Colima, Coahuila, Baja California Sur, and Querétaro are leading in FinTech usage, exhibiting significantly higher adoption rates than the national average. This indicates that these states are more advanced in their integration of digital financial services. In contrast, Chiapas and Puebla show considerably lower levels of FinTech adoption, falling behind the rest of the country.

The confidence intervals provide insight into the reliability of these findings with narrower ranges indicating more precise measurements of FinTech usage within each state. Similarly, the standard error reflects the uncertainty inherent in these measurements with smaller standard errors implying greater confidence in the results.

Table 1 Application of the Indicator in the 32 States of Mexico

State	Estimation index	Standard error	Confidence interval 95%
Colima	5.26	0.21	[4.85, 5.67]
Coahuila de Zaragoza	5.11	0.20	[4.72, 5.50]
Baja California Sur	5.10	0.19	[4.72, 5.48]
Querétaro	5.06	0.21	[4.66, 5.47]
Chihuahua	5.05	0.19	[4.68, 5.42]
Yucatán	5.03	0.19	[4.65, 5.41]
Quintana Roo	5.02	0.20	[4.63, 5.41]
Jalisco	4.98	0.21	[4.57, 5.40]
Tabasco	4.88	0.20	[4.48, 5.28]
Campeche	4.82	0.23	[4.36, 5.27]
Nayarit	4.76	0.23	[4.30, 5.22]
Baja California	4.74	0.19	[4.38, 5.11]
Nuevo León	4.54	0.19	[4.17, 4.91]
Tamaulipas	4.36	0.20	[3.97, 4.75]
Veracruz	4.33	0.24	[3.86, 4.80]
Sinaloa	4.27	0.25	[3.79, 4.76]
Durango	4.27	0.24	[3.80, 4.74]
CDMX	4.25	0.21	[3.83, 4.67]
Aguascalientes	4.23	0.23	[3.79, 4.67]
Morelos	4.16	0.26	[3.65, 4.67]
Michoacán de Ocampo	4.15	0.25	[3.67, 4.63]
Sonora	4.10	0.22	[3.67, 4.53]
México	4.02	0.21	[3.62, 4.42]
Tlaxcala	3.97	0.17	[3.63, 4.31]
San Luis Potosí	3.93	0.23	[3.47, 4.38]
Oaxaca	3.84	0.29	[3.28, 4.41]
Guerrero	3.70	0.23	[3.26, 4.15]
Hidalgo	3.70	0.19	[3.32, 4.08]
Zacatecas	3.69	0.24	[3.22, 4.15]
Guanajuato	3.62	0.20	[3.23, 4.02]
Chiapas	3.43	0.30	[2.85, 4.02]
Puebla	3.43	0.20	[3.05, 3.82]
National average	4.31	0.05	[4.21, 4.41]

Source: Authors' own elaboration with data from ENDUTIH 2022, INEGI

Overall, the analysis highlights significant disparities in the utilization of financial technology across Mexico. While some regions show a strong preference for digital financial solutions, others remain less engaged, emphasizing the need for targeted interventions to promote FinTech adoption in underperforming areas.

Table 2 presents the percentage of individuals in each Mexican state who do not use online payments and/or have not installed mobile banking applications. This data underscores the disparities in digital financial inclusion across the country.

Table 2 Percentage of people that does not payment by internet and install banking mobile

State	No payments by internet (%)	No installation applications mobile (%)
Colima	46.09	48.27
Coahuila de Zaragoza	46.43	50.55
Baja California Sur	33.92	59.11
Querétaro	46.82	51.04
Chihuahua	43.71	53.35
Yucatán	49.68	49.73
Quintana Roo	52.79	47.79
Jalisco	47.13	52.19
Tabasco	61.66	44.23
Campeche	44.89	56.48
Nayarit	49.88	54.10
Baja California	46.04	56.89
Nuevo León	43.13	62.27
Tamaulipas	46.49	63.01
Veracruz de Ignacio de la Llave	54.37	58.20
Sinaloa	39.71	68.95
Durango	42.63	67.13
Ciudad de México	40.49	68.86
Aguascalientes	50.18	62.72
Morelos	49.39	64.40
Michoacán de Ocampo	56.18	60.00
Sonora	36.79	73.78
México	55.33	62.78
Tlaxcala	63.55	58.12
San Luis Potosí	51.52	66.89
Oaxaca	56.20	65.12
Guerrero	62.27	63.42
Hidalgo	57.52	66.63
Zacatecas	52.87	70.00
Guanajuato	56.02	68.96
Chiapas	59.58	69.70
Puebla	64.80	66.24
National average	50.95	60.82

Source: Authors' own elaboration with data from ENDUTIH 2022, INEGI

This data highlights the uneven distribution of digital financial services across Mexican states, emphasizing the need for a nuanced approach to measuring FinTech usage within the proposed index. The data reveals a significant difference between the prevalence of internet payments and mobile banking app installations. While both are valuable indicators, the consistently higher rates of non-adoption of mobile banking across several states suggest a stronger relationship between mobile

Table 3 Reasons because people do not do payments by internet

State	Do not trust security	No credit card	Not interested	Not allowed	No knowledge	Not old enough	Other reasons	Prefers in-person
Colima	10.8%	4.0%	11.0%	0.7%	15.5%	0.0%	0.4%	57.7%
Coahuila de Zaragoza	12.4%	1.8%	9.6%	0.7%	11.1%	0.0%	0.4%	64.0%
Baja California Sur	10.9%	2.5%	20.0%	0.0%	11.7%	0.1%	0.4%	54.4%
Querétaro	8.6%	4.6%	13.0%	0.1%	11.6%	0.4%	1.8%	60.0%
Chihuahua	17.0%	5.2%	15.1%	1.1%	15.8%	0.6%	0.4%	44.7%
Yucatán	17.4%	4.5%	12.7%	0.1%	15.4%	0.1%	2.1%	47.7%
Quintana Roo	14.9%	5.2%	8.6%	0.6%	8.5%	0.0%	1.7%	60.5%
Jalisco	9.3%	4.2%	17.0%	0.8%	11.0%	0.0%	0.9%	56.7%
Tabasco	9.6%	3.7%	26.9%	0.7%	14.8%	0.0%	1.3%	43.0%
Campeche	11.1%	5.0%	15.5%	0.4%	14.5%	0.0%	1.8%	51.8%
Nayarit	14.6%	4.4%	15.8%	0.7%	13.7%	0.2%	0.9%	49.8%
Baja California	13.5%	2.3%	14.0%	0.9%	6.2%	0.0%	1.6%	61.7%
Nuevo León	9.0%	2.8%	16.5%	1.0%	8.9%	0.0%	0.9%	60.9%
Tamaulipas	13.5%	3.2%	10.6%	0.2%	14.7%	0.0%	1.2%	56.8%
Veracruz de Ignacio de la Llave	9.1%	4.9%	16.8%	0.8%	14.3%	0.0%	0.5%	53.7%
Sinaloa	7.2%	3.5%	21.9%	0.5%	18.0%	0.0%	1.0%	48.0%
Durango	11.7%	6.4%	25.2%	0.5%	14.0%	0.4%	2.8%	39.0%
Ciudad de México	10.3%	5.2%	12.3%	1.4%	5.3%	0.3%	1.9%	63.2%

(continued)

Table 3 (continued)

State	Do not trust security	No credit card	Not interested	Not allowed	No knowledge	Not old enough	Other reasons	Prefers in-person
Aguascalientes	9.5%	2.4%	22.0%	0.6%	10.7%	0.0%	0.7%	54.2%
Morelos	13.4%	6.6%	26.2%	0.6%	9.8%	0.0%	0.2%	43.2%
Michoacán de Ocampo	11.1%	5.0%	28.9%	0.2%	14.3%	0.0%	0.7%	39.8%
Sonora	12.6%	6.3%	11.9%	0.4%	12.7%	0.2%	0.8%	55.1%
México	16.3%	4.9%	17.1%	0.5%	9.1%	0.1%	0.3%	51.7%
Tlaxcala	20.3%	7.5%	18.5%	0.1%	12.5%	0.0%	0.9%	40.2%
San Luis Potosí	13.3%	5.1%	18.0%	0.0%	9.2%	0.0%	1.3%	53.1%
Oaxaca	11.3%	8.6%	20.7%	0.2%	16.3%	0.2%	1.3%	41.4%
Guerrero	14.6%	6.0%	18.7%	0.5%	21.5%	0.0%	1.6%	37.1%
Hidalgo	11.4%	6.7%	15.8%	0.8%	10.1%	0.2%	0.3%	54.7%
Zacatecas	10.5%	5.1%	25.4%	0.5%	14.1%	0.0%	1.1%	43.4%
Guanajuato	15.1%	3.3%	20.3%	0.1%	8.1%	0.2%	0.1%	52.8%
Chiapas	9.2%	4.9%	25.3%	0.0%	14.8%	0.3%	0.7%	44.9%
Puebla	19.8%	5.7%	17.0%	0.1%	9.5%	0.4%	0.9%	46.6%
National average	12.7%	4.7%	17.4%	0.6%	11.3%	0.1%	0.9%	52.3%

Source: Authors' own elaboration with data from ENDUTIH 2022, INEGI

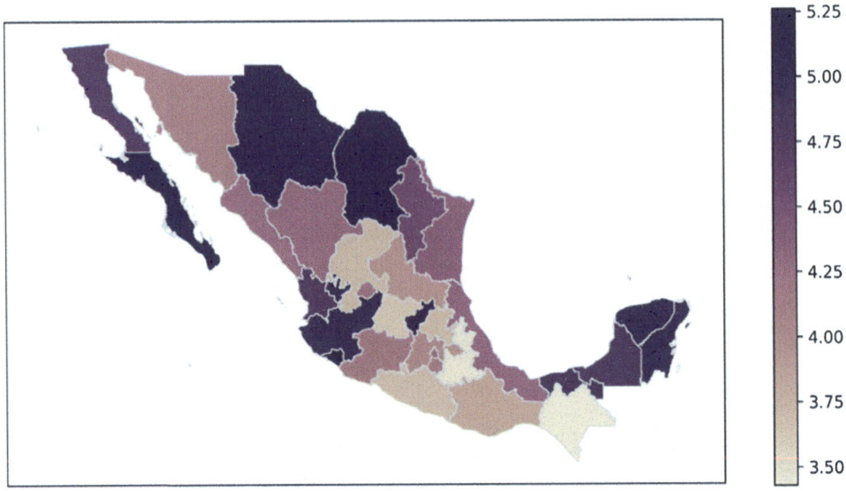

Fig. 1 Levels of access to FinTech across the federal entities, including Ciudad de México. Source: Authors' own elaboration, with the use of Dive Into Python 3 software [18]

banking and broader FinTech penetration. This discrepancy is likely due to factors such as digital literacy, which can hinder the adoption of mobile banking, while internet payments often prove more accessible and user-friendly.

Considering this data, the proposed FinTech usage index assigns a higher weighting to mobile banking app installation. This weighting reflects the stronger correlation observed between mobile banking adoption and overall FinTech usage as shown in Table 2. This approach ensures a more accurate representation of the current landscape and facilitates the development of policies aimed at improving financial inclusion and expanding the use of financial technologies throughout Mexico.

Additionally, further data supports our analysis of FinTech usage in Mexico. Table 3 details the reasons why individuals in different Mexican states do not make online payments. This breakdown provides valuable insights into the barriers preventing wider adoption of digital financial services.

The data in Table 3 highlights an intriguing dynamic in Mexico's financial landscape. Ciudad de México, despite being the largest financial center with a substantial population and extensive access to traditional financial resources, exhibits a contrasting trend in FinTech utilization compared to other states. This trend suggests that while Ciudad de México benefits from a well-established banking infrastructure, residents in other states may display greater trust and confidence in FinTech solutions. Interestingly, despite being a major financial hub, Ciudad de México shows a lower rate of FinTech adoption compared to other states. This may be attributed to its established traditional banking infrastructure and varying levels of trust in FinTech solutions. Reasons for not using online payments vary across states with preference to do it in person, not interest or do not need it and lack of trust in security to do it, this analysis shows that limited access to credit cards is not one of the notable barriers. The data underscores significant regional disparities in financial technology adoption in Mexico. Colima, Coahuila, Baja California Sur, and Querétaro lead in FinTech usage, whereas Chiapas and Puebla lag far behind. The analysis reveals a strong correlation between mobile banking app adoption and broader FinTech penetration, emphasizing the need for nuanced approaches to measuring FinTech usage and targeted policies to improve financial inclusion. Finally, the following map visually illustrates the levels of access to FinTech across the federal entities, including Ciudad de México (Fig. 1).

5 Conclusions

This research makes a significant contribution to the understanding of financial technology (FinTech) adoption in Mexico. By developing a proposal to measure the use of financial technology through the FinTech Usage Index based on readily available data from the ENDUTIH survey, as this study provides a practical and replicable tool for assessing FinTech penetration across different regions of the country. The index's clear methodology and use of accessible data enhance its value for policymakers, researchers, and other stakeholders interested in monitoring and promoting financial inclusion. The findings highlight substantial regional disparities in FinTech usage which are critical considerations for targeted policy interventions.

The analysis reveals significant variations in FinTech adoption levels across Mexican states, with some regions demonstrating considerably higher usage rates than others. This uneven distribution underscores the need for tailored strategies to address the digital divide and promote access to digital financial services. Identifying these regional disparities is a crucial first step in designing effective policies to encourage broader FinTech adoption and improve financial inclusion nationwide. The study's results offer valuable insights for policymakers seeking to promote inclusive growth and economic empowerment.

Furthermore, the study explores the relationship between FinTech usage and several key socio-economic factors. While a detailed presentation of the results from the statistical model is available in a preprinted article. The analysis emphasizes the importance of these factors when designing and implementing policies aimed at promoting wider FinTech adoption. Understanding the influence of these variables on FinTech usage is essential for developing targeted, effective, and equitable interventions. This approach ensures that policy decisions are data-driven and tailored to the specific needs of diverse population segments.

Future research will build upon this study in several important ways. A time-series analysis leveraging data from previous years of the ENDUTIH survey will be conducted to track changes in FinTech adoption over time, thus revealing trends and patterns in the evolution of digital financial inclusion in Mexico. Moreover, advanced statistical techniques such as the X-model [21], could be employed to refine the analysis and more accurately assess the impact of socio-economic factors on FinTech adoption. In addition to these quantitative analyses, a qualitative research will explore user experiences and challenges associated with FinTech adoption in Mexico. This multifaceted approach will provide a comprehensive and nuanced understanding of FinTech usage and its impact on Mexican society.

References

1. Arredondo, V., Martínez-Panero, M., Peña, T., et al.: Mathematical political districting taking care of minority groups. Annals Oper. Res. **305**, 375–402 (2021)
2. Arredondo, V., Falcon, V., Lashkari, Z.: Bridging the digital divide: Examining socioeconomic factors influencing Fintech usage in Mexico (2025). Preprint
3. Carbó-Valverde, S., Cuadros-Solas, P., Rodríguez-Fernández, F.: Financial inclusion, digitalization, and the measurement of financial services access. J. Bank. Finance **72**(1), 34–45 (2016)
4. Fi, L.: La evolución de las fintech: Transformando el mundo financiero. CDETECH (2023). https://cdetech.org/la-evolucion-de-las-fintech-transformando-el-mundo-financiero/
5. Chen, S., Dai, X., Kong, D., Li, Y.: Digital financial inclusion and economic growth: A cross-country analysis. J. Financ. Stab. **48**, 100762 (2020)
6. Demirgüç-Kunt, A., Klapper, L., Singer, D., Van Oudheusden, P.: The global Findex database 2014: Measuring financial inclusion around the world. Technical report, World Bank Group (2014)
7. Forbes: Historia del fintech: Revolucionando las finanzas (2024)
8. Greene, W.H.: Econometric Analysis, 5th edn. Prentice Hall, Upper Saddle River (2002)
9. Zaltzman, L.: Historia del fintech: Revolucionando las finanzas. RappiCard (2024). https://rappicard.mx/2024/04/17/historia-del-fintech/
10. IBM Corp: IBM SPSS Statistics for Windows, Version 28.0. Armonk, New York (2023)
11. Instituto Nacional de Estadística y Geografía (INEGI) and Instituto Federal de Telecomunicaciones (IFT): Encuesta nacional sobre disponibilidad y uso de tecnologías de la información en los hogares (endutih) 2022. Technical report, INEGI & IFT (2023)
12. Instituto Nacional de Estadística y Geografía (INEGI) and Instituto Federal de Telecomunicaciones (IFT): Marco geo-estadístico. Technical report, INEGI & IFT (2024)
13. Klapper, L., Love, I.: Financial inclusion and Fintech impact. J. Finance **65**(3), 123–135 (2011)
14. Microsoft Corporation: Visual Studio Code (Versión 1.88) [Entorno de desarrollo integrado] (2024)
15. Mishkin, F.S., Eakins, S.G.: Financial Markets and Institutions, 8th edn. Pearson, New Delhi (2018)
16. Moawed, S.A., Abd El-Aziz, A.H.: The estimation and interpretation of ordered logit models for assessing the factors connected with the productivity of Holstein–Friesian dairy cows in Egypt. Trop. Anim. Health Prod. **54**(6), 345 (2022)
17. Montero Granados, R.: Análisis de regresión para variables categóricas: Teoría y práctica. Ediciones Universitarias, Barcelona (2016)
18. Pilgrim, M., Willison, S.: Dive into Python 3, vol. 2. Springer, New York (2009). Version 3.12.2
19. RStudio Team: RStudio: Integrated Development for R (2020). Version 2023.12.1+402
20. Vara Rey, F.J.: Modelos estadísticos avanzados para datos categóricos: Aplicaciones en ciencias sociales y económicas. Editorial Complutense, Madrid (2016)
21. Vásquez, A.R., Pérez, G.E., Núñez-Antonio, G., Urbina, J.U.M.: La replicabilidad en la ciencia y el papel transformador de la metodología estadística de knockoffs. Sahuarus. Rev. Electrón. Mat. **8**(1), 1–22 (2024)
22. Zigurat: The evolution of Fintech: From Fintech 1.0 to Fintech 4.0, August 25 (2022). Accessed 11 Dec 2024

Some Theoretical Foundation for Protein Identification Through Cyclic Codes

Diana H. Bueno-Carreño, Mauricio Frieri, and Mario E. Duarte

Abstract In recent work, the potential relationship between protein synthesis processes and digital transmission systems has been investigated. Progress in this matter could lead to conclusions about the evolution and connection between species. An intersection between Coding Theory and biological systems lies in the representation of biological sequences as code words of BCH codes. In the article *Construction of Cyclic Codes over* $\mathbb{Z}_{20}$ *for Identifying Proteins* (Galíndez and Duarte, Construction of cyclic codes over $\mathbb{Z}_{20}$ for identifying proteins, pp. 40–51, 2019), three algorithms were designed to identify odd-length polypeptide sequences as code words of cyclic codes. However, a formal foundation of the algorithms was not provided. This work establishes a mathematical foundation for all three algorithms, proposes a novel approach to improve the performance of the protein identification process; and presents some examples to demonstrate its application.

Keywords Cyclic codes · Ring extension · Protein identification

1 Introduction

In Coding Theory, codes with algebraic structures are designed to codify messages in a manner that the receptor detects and, in some cases, corrects alterations that might appear during the message transmission. A similar phenomenon occurs in

D. H. Bueno-Carreño (✉)
Pontificia Universidad Javeriana, Cali, Colombia
e-mail: dhbueno@javerianacali.edu.co

M. Frieri
Universiteit Leiden, Leiden, Netherlands
e-mail: mauriciofrieri@gmail.com

M. E. Duarte
DyG Representaciones SAS Bogotá, Bogotá, Colombia
e-mail: mduarte@dyg.com.co

P. E. Harris et al. (eds.), *Women in Mathematics in Latin America*,
Research Perspectives Ghent Analysis and PDE Center 14,
https://doi.org/10.1007/978-3-031-97857-9_4

DNA translation and transcription steps, where the DNA genetic information is transformed to synthesize proteins. In this scenario, the biological information is transmitted from the DNA to RNA and polypeptide sequences (proteins); this is called the *central dogma of molecular biology*. Several researchers have made important advances to demonstrate that an error-correcting code structure lies within the biological information transmission system; based on the fact that proteins undergo alterations at the time of their synthesis, but continue to fulfill the same functions.

In the paper *Construction of Cyclic Codes over* $\mathbb{Z}_{20}$ *for Identifying Proteins* [4], three algorithms were designed to identify odd-length polypeptide sequences as code-words of cyclic codes. However, these algorithms lack an explicit mathematical theoretical foundation. Algorithm 1, given an integer number (length of the code: n), computes the corresponding cyclotomic cosets and the factorization of the polynomial $x^n - 1$. Algorithm 2, given an integer l and a polynomial $p(x)$, efficiently finds a computational representation of the element $\alpha^l \mod p(\alpha)$. Algorithm 3, given a target polynomial $s(x)$, retrieves the generator polynomial $g(x)$ of the BCH code with the largest minimum distance such that $s(x)$ belongs to the code.

In this work, we present a mathematical foundation of the algorithms of this article, accompanied by an example describing how to apply Algorithm 1; and we introduce a novel proposition (BDF proposition) for improving the efficiency of Algorithm 1. It should be mentioned that making the process more efficient means that the search for mathematical structures in biology can be carried out under better conditions.

This paper is organized as follows. Section 2 presents the general aspects of the theoretical background necessary to understand the protein identification methodology proposed in the article [4]. Section 3 presents the theoretical foundation corresponding to each algorithm, along with the BDF proposition and some examples. Finally, conclusions are drawn in Sect. 4.

2 Theoretical Foundation

In [4], three algorithms were developed to identify proteins as code-words of a cyclic code over $\mathbb{Z}_{20}$ using, primarily, results from [8] and [5].

The first step consists of identifying a succession of amino acids as a sequence over $\mathbb{Z}_{20}$. For achieving this, [4] uses the Dayhoff's matrix, which is a manner of representing amino acids in a circular topology that arranges them in such a way that amino acids more likely to swap positions correspond to close residues [2].

The sequence over $\mathbb{Z}_{20}$ (s_{20}) is represented as a polynomial $s(x)$ over $\mathbb{Z}_{20}[x]$. The intention now is to construct an error-correcting code C_{20} such that $s(x) \in C_{20}$. The strategy of [4] consists of using the Chinese Remainder Theorem to factorize $s(x)$ into two sequences $s_4(x)$ and $s_5(x)$ and construct codes C_4 and C_5 over $\mathbb{Z}_4$ and $\mathbb{Z}_5$, respectively, such that $s_4(x) \in C_4$ and $s_5(x) \in C_5$. This factorization is possible thanks to the isomorphism between $\mathbb{Z}_{20}$ and $\mathbb{Z}_4 \oplus \mathbb{Z}_5$. Finally, the codes C_4

and C_5 are the base parts of C_{20}, which is obtained by using the formerly mentioned isomorphism. We recommend section 2.2 of [4] for further information about this process.

The problem is reduced to the construction of C_4 and C_5, such that $s_4(x)$ and $s_5(x)$ belong to the corresponding codes. In [4], the construction of C_5 over the field $\mathbb{Z}_5$ follows the well-known approach outlined in [11], while the construction of C_4 over the ring $\mathbb{Z}_4$, is detailed in the following paragraphs.

To construct BCH codes over $\mathbb{Z}_4$, in the literature, we find articles addressing polynomial factorization, construction of minimal polynomials, and estimation of the BCH bound for codes over finite commutative rings in [1] and over chains of commutative rings in [7]. In addition, in [8] Shankar explains how to construct BCH codes of length n over $\mathbb{Z}_M$, with $M = \prod_{i=1}^{l} p_i^{k_i}$ and $\text{mcd}(n, M) = 1$, where each p_i is a prime and each k_i is a natural number; based on the factorization of $x^n - 1$ over the unit group of one of the extensions of $\mathbb{Z}_M$. Such extension is given by:

$$R = \frac{\mathbb{Z}_{p^k}[y]}{\Phi(y)},$$

where $\mathbb{Z}_{p^k}[y]$ is the polynomial ring in the indeterminate y over $\mathbb{Z}_{p^k}$ and $\Phi(y)$ is a polynomial of degree r, irreducible over $\mathbb{Z}_p$ (it is also irreducible over $\mathbb{Z}_{p^k}$). R is also denoted as $GR(p^k, r)$ and is called *Galois extension* of $\mathbb{Z}_{p^k}$.

In [4], the utilized procedure for factorizing $x^n - 1$ over $\mathbb{Z}_4$ consists of two steps: (1) factorize $x^n - 1$ over $\mathbb{Z}_2$ [8], and (2) apply the process exposed by Huffman and Pless in [5, p. 479] to all minimal polynomial obtained from the prior step. This process relies on the Graeffe's method, adapted to linear codes over $\mathbb{Z}_4$ in [9, Sect. 6, Theorem 2] and [10], to apply the Hensel's Lemma [5, Theorem 12.3.4]. This lemma can be applied using Graeffe's method as follows [5, p. 479]:

I. Given $h(x)$ an irreducible factor of $x^n + 1$ in $\mathbb{Z}_2[x]$ we rewrite it as $h(x) = e(x) + o(x)$, where $e(x)$ is the sum of the terms of even degree of $h(x)$ and $o(x)$ is the sum of the terms of odd degree of $h(x)$.
II. We construct $g(x)$, an irreducible polynomial in $\mathbb{Z}_4[x]$, which is equal to $h(x)$ mod 2 and verifies that $g(x^2) = \pm\left(e(x)^2 - o(x)^2\right)$.

The factorization of $x^n - 1$ over $\mathbb{Z}_4$ is unique when n is odd according to Theorem 12.3.7 of [5].

Next, some results and definitions are presented to understand the operation of Algorithm 1 of [4]. Let R^* be the set of units of R.

Definition 1 (q-Cyclotomic Coset) [5, p. 114] The q-cyclotomic cosets of s mod n is the set:

$$\mathfrak{C}_s = \left\{s, sq, sq^2, \ldots, sq^{r-1}\right\},$$

where r is the smallest positive integer such that $sq^r \equiv s \bmod n$.

Theorem 1 ([8], Theorem 2) *R^* contains exactly one cyclic subgroup with order co-prime to p. This cyclic subgroup has order $p^r - 1$.*

Theorem 2 ([8], Theorem 3) *Let f be a generator of a cyclic subgroup of order n in R^* where $mcd(n, p) = 1$. Then, the polynomial $x^n - 1$ can be factorized as $x^n - 1 = (x - f)(x - f^2) \cdots (x - f^n)$ if and only if $R_p(f)$ has order n in K^*, where $R_p(f)$ is the least nonnegative remainder when f is divided by p.*

In Theorem 1, Shankar states that R^* has only one cyclic subgroup of order co-prime to p, and its order is $p^r - 1$. This is not correct and probably is a redaction error. However, the largest cyclic subgroup with order co-prime to p indeed has order $p^r - 1$. This follows from the arguments exposed in [6]. Fortunately, this detail doesn't affect the validity of the other results shown in [8] and [4].

3 Results

In this section, we establish the theoretical foundations for the algorithms outlined in [4], introduce the novel BDF proposition, and provide examples.

3.1 Algorithm 1: Factorization of $x^n - 1$

Algorithm 1 uses a database with irreducible polynomials over $\mathbb{F}_p$ of various degrees and several values of n and factorizes $x^n - 1$ over $\mathbb{F}_p$ (Fig. 1). It is worth mentioning that, in the case of $\mathbb{Z}_4$, first Algorithm 1 is executed, factoring $x^n - 1$ over $\mathbb{Z}_2 \cong \mathbb{F}_2$ and then the Graeffe's method is applied to obtain the factorization over $\mathbb{Z}_4$ as in [5].

Algorithm 1: Factorization of $x^n - 1$ over $\mathbb{F}_p$

On input: Data base, $p(x)$, n
1: Find l and r such that $l \cdot n = p^r - 1$.
2: Get the primitive polynomial $p(x)$ of degree r form the data base.
3: Compute $\beta = \alpha^l$.
4: Construct the multiplicative subgroup $\mathbf{H}_n$ of order n
$\mathbf{H}_n =< \beta > \pmod{p(\alpha)}$
5: Find the cyclotomic cosets: $CL = \{cl_p(i, n) : i \geq 0\}$, where $cl_p(i, n) := \{i \cdot p^j \pmod{n} : j \geq 0\}$.
6: **For** Every cyclotomic coset $cl_p(i, n)$ in CL **do**
7: Compute the minimal polynomial:
$f_i(x) = \prod_{j \in cl_p(i,n)} (x - \beta^j)$.
Then, $f_i(x)$ is a minimal polynomial over $\mathbb{F}_q$ that satisfies $f_i(x) | x^n - 1$.
8: **End for**
9: **Return** CL, $F_n = \{f_i(x) : i \geq 0\}$.

Fig. 1 Algorithm 1 rewritten from [4]

In step 1, integer values for l and r are computed such that $l \cdot n = p^r - 1$. This condition is required because of the interaction between Theorems 1 and 2: To apply Theorem 2 we must find an element f in R^* that generates a multiplicative subgroup of order n, where $mcd(n, p) = 1$. According to Theorem 1, the largest cyclic subgroup of order relative prime to p has order $p^r - 1$ and all other groups with said constriction are subgroups of this one. Hence, with this step, we obtain values of r and l that allow us to construct a convenient Galois extension and a subgroup of said extension.

In step 2 we chose a primitive polynomial $p(x)$ of degree r that will allow us to construct R. Then, in step 3, we use Algorithm 2 of [4] to compute $\beta = \alpha^l$, where α is a primitive root of $p(x)$. Thus, β is a generator of a cyclic subgroup of order n that will be helpful to factor $x^n - 1$. In step 4 we generate the subgroup, which is denoted by $\mathbf{H}_n$.

In step 5 the p-cyclotomic cosets mod n are found. Steps 6, 7, and 8 correspond to a *for* cycle that calculates the minimal polynomials that factor $x^n - 1$ by grouping the powers of β with exponents from the same cyclotomic coset in linear factors. Finally, step 9 returns the cyclotomic cosets and the minimal polynomials. Note that, within the algorithm, *CL* stands for the cyclotomic cosets, while here they are denoted as $\mathfrak{C}_s$.

Next, we present a step-by-step example to demonstrate how Algorithm 1 works.

3.1.1 Example 1: Factorization of $x^{21} - 1$ over $\mathbb{Z}_4$

We chose $l = 3$ and $r = 6$ since $3 \cdot 21 = 2^6 - 1 = 63$. We also chose $p(x) = x^6 + x + 1$ primitive over $\mathbb{Z}_2$ and of degree r. Using Algorithm 2, we calculate $\beta = \alpha^l$ which is α^3 and now we compute $\mathbf{H}_{21}$.

The 2-cyclotomic cosets mod 21 are: $\mathfrak{C}_0 = \{0\}$, $\mathfrak{C}_1 = \{1, 2, 4, 8, 16, 11\}$, $\mathfrak{C}_3 = \{3, 6, 12\}$, $\mathfrak{C}_5 = \{5, 10, 20, 19, 17, 13\}$, $\mathfrak{C}_7 = \{7, 14\}$ and $\mathfrak{C}_9 = \{9, 18, 15\}$. We proceed to calculate the minimal polynomial for each coset:

$$f_k(x) = \prod_{j \in \mathfrak{C}_k} (x - \beta^j). \tag{1}$$

It can be seen that the minimal polynomials are the following:

$$\begin{aligned} f_1(x) &= x^6 + x^4 + x^2 + x + 1 \\ f_3(x) &= x^3 + x^2 + 1 \\ f_5(x) &= x^6 + x^5 + x^4 + x^2 + 1 \\ f_7(x) &= x^2 + x + 1 \\ f_9(x) &= x^3 + x + 1. \end{aligned} \tag{2}$$

And, after applying Graeffe's method they become:

$$\begin{aligned}\overline{f_1}(x) &= x^6 + 2x^5 + 3x^4 + 3x^2 + x + 1\\ \overline{f_3}(x) &= x^3 + 3x^2 + 2x - 1\\ \overline{f_5}(x) &= x^6 + x^5 + 3x^4 + 3x^2 + 2x + 1\\ \overline{f_7}(x) &= x^2 + x + 1\\ \overline{f_9}(x) &= x^3 + 2x^2 + x - 1.\end{aligned} \tag{3}$$

The following lemma gives way to Proposition BDF (named after the initials of the author's surnames), a new result that is stated as an improvement proposal for Algorithm 1:

Lemma 1 ([5, Lemma 4.4.7]) *Let $f(x) \in \mathbb{F}_q[x]$. If $f(x) = f_0 + f_1 x + \cdots + f_a x^a$, we define the reciprocal polynomial of f as $f^*(x) = x^a f(x^{-1}) = f_a + f_{a-1}x + \cdots + f_1 x^{a-1} + f_0 x^a$.*

(i) If $\beta_1, \ldots \beta_r$ are the non zero roots of $f(x)$ in some extension of $\mathbb{F}_q$, then $\beta_1^{-1}, \ldots \beta_r^{-1}$ are the non zero roots of $f^*(x)$ in said extension.

(ii) If $f(x)$ is irreducible over $\mathbb{F}_q$, so is $f^*(x)$.

(iii) If $f(x)$ is a primitive polynomial, $f^*(x)$ is also a primitive polynomial.

Proposition 3 (Proposition BDF) *The number of minimal polynomials that must be calculated in step 7 of Algorithm 1 lays in the interval $[\gamma/2, \gamma] \bigcap \mathbb{Z}$, where γ is the number of q-cyclotomic cosets modulo n.*

Proof Thanks to Lemma 1 we know that when calculating the minimal polynomial of the elements of the q-cyclotomic coset of k, we can immediately infer the minimal polynomial of the q-cyclotomic coset of $-k$. The worst case scenario will be when k and $-k$ belong to the same q-cyclotomic coset for all k; hence, being necessary to calculate all the γ different minimal polynomials, while in the best case scenario, it will only be necessary to calculate $\gamma/2$ minimal polynomials. □

The worst-case scenario corresponds to linear codes with complementary-duals (LCD) which have many applications in cryptography, communications, and data storage [3].

To better understand the Proposition BDF, we present two visual examples. In Fig. 2, we want to factor $x^9 - 1$ over $\mathbb{Z}_2$. The 2-cyclotomic cosets mod 9 are: $\mathfrak{C}_0 = \{0\}$, $\mathfrak{C}_1 = \{1, 2, 4, 5, 7, 8\}$ and $\mathfrak{C}_3 = \{3, 6\}$. We visualize the 9th-roots of unity in the typical circular manner and observe a symmetry related to the 2-cyclotomic cosets and its relation to Proposition BDF.

We can label each root of unity according to their exponent as the cyclic group $\mathbf{H}_n$. Each cyclotomic coset is colored with a different color. The effect of reflection, using the dashed vertical line as a pivot, is that the kth element is sent to the $-k$th

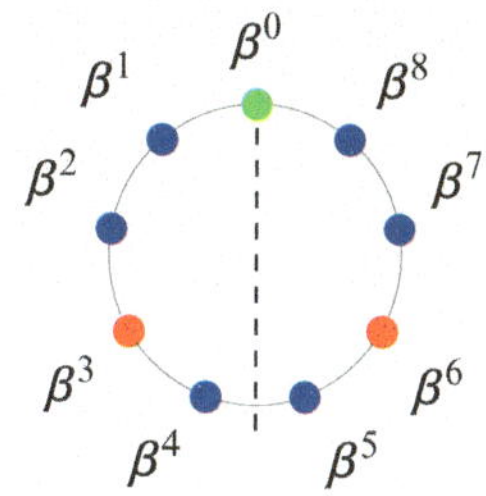

Fig. 2 Visual representation for the case $x^9 - 1$

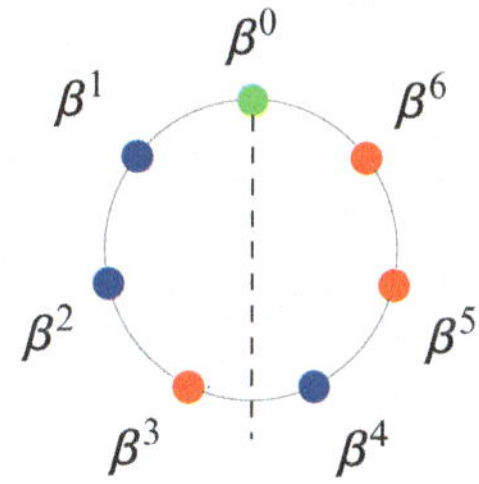

Fig. 3 Visual representation for the case $x^7 - 1$

element. Note that this reflection preserves coloring (i.e. blue goes to blue, and so on), meaning that all the minimal polynomials of the cyclotomic cosets are self-reciprocal. This is an example of the worst-case scenario.

Consider the same visual representation for the case of $n = 7$ in Fig. 3.

In this case, we can observe that the reflection interchanges the color, meaning that the reciprocal minimal polynomial of each cyclotomic coset (excluding $\mathfrak{C}_0$) is a different minimal polynomial in the factorization of $x^n - 1$. The same diagrams can be constructed for $p = 5$.

As explained before, the savings in the number of calculations depends on n. For a given word length n, we can estimate the improvement that the BDF proposition provides by a simple calculation regarding the number of cyclotomic cosets and those that are not self-reciprocal. To have a relevant estimation to real-world applications, we computed these estimates for $n \in [31, 93]$, for $p = 2$ and $p = 5$ since 31, 63 and 93 are the values used in [4] to inspect the validity of their algorithms. These values were selected based on two criteria: they are not multiples of 2 or 5 and proteins with these lengths have been found. Figures 4 and 5 show said improvements for $p = 2$ and $p = 5$ respectively.

For $p = 2$ the BDF proposition reduces the number of calculated minimal polynomials by 25%, and for $p = 5$ by 21.4%. If the range is expanded, the improvement slightly increases.

3.2 *Algorithm 2: Computing α^l Modulo $p(\alpha)$*

The Algorithm 2 efficiently computes the value of α^l modulo $p(\alpha)$, given the values of l, p and the evaluated polynomial in α, $p(\alpha)$ (Fig. 6).

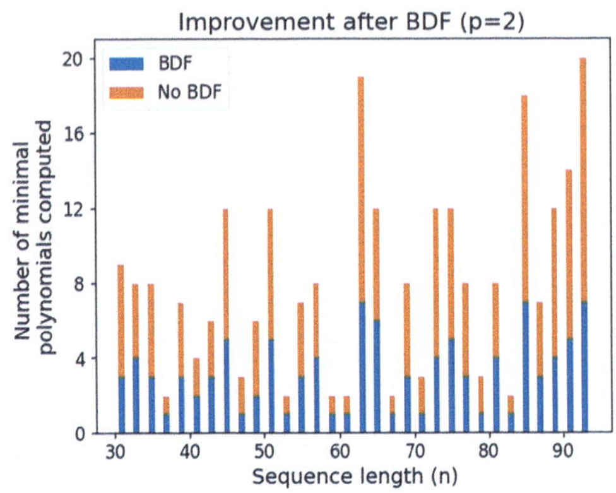

Fig. 4 BDF improvement $p = 2$

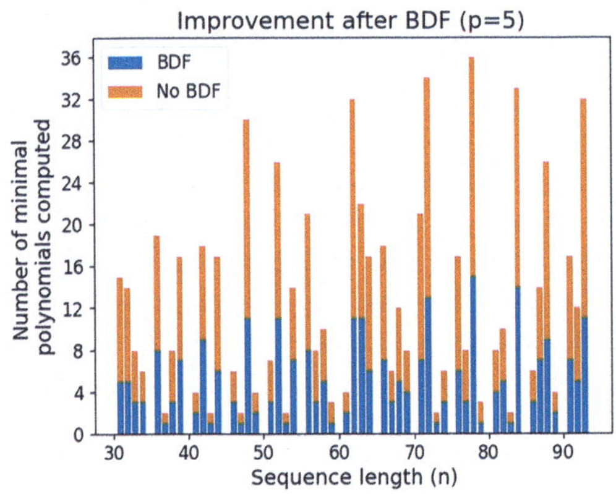

Fig. 5 BDF improvement $p = 5$

Algorithm 2: $\alpha^l \pmod{p(\alpha)}$

On input: l, p, $p(\alpha)$.

1: Define r as the degree of $p(\alpha)$.

2: **If** $l < p \cdot r$ **Then**

3: **Return** Remainder of $\frac{\alpha^l}{p(\alpha)}$ (low cost operation).

4: **End if**

5: Find l_1 and l_2, such that: $l = l_2 + (l_1)(p^m)$, $l_1 < p \cdot r$, and $m \leq \left\lfloor \frac{\log l}{\log p} \right\rfloor$, and hence, $\alpha^l = \alpha^{l_2}\left(\alpha^{l_1}\right)^{p^m}$

6: Get the remainder of $\frac{\alpha^{l_1}}{p(\alpha)}$ (low cost operation) i.e. $r(\alpha) = \Sigma_{i=0}^{r-1} a_i \alpha^i$.

7: Apply the equality: $r(\alpha)^{p^m} = r(\alpha^{p^m})$, *i.e.*

$$\left(\sum_{i=0}^{r-1} a_i \alpha^i\right)^{p^m} = \sum_{i=0}^{r-1} a_i \alpha^{i \cdot p^m}$$

8: **For** each i such that $a_i \neq 0$ do

9: Compute $\alpha^{(i+l_2)\cdot p^m} \pmod{p(\alpha)}$ (calling again this algorithm).

10: **End for**

11: **Return** $\Sigma_{i=0}^{r-1} a_i \left(\alpha^{(i+l_2)\cdot p^m}\right) \pmod{p(\alpha)}$ (low cost operation because of step 9).

Fig. 6 Algorithm 2: Computing α^l modulo $p(\alpha)$

We present the following result that will be useful in this section:

Definition 2 (Frobenius Automorphism) [5, p. 112] Let $\mathbb{F}_q$ be the finite field of q elements and let p be the characteristic of $\mathbb{F}_q$. We define $\sigma_p : \mathbb{F}_q \to \mathbb{F}_q$ as $\sigma_p(\alpha) = \alpha^p$ for all $\alpha \in \mathbb{F}_q$.

This algorithm is invoked by Algorithm 1, and its purpose is to make the process of factorizing $x^n - 1$ efficient. For this, the Frobenius automorphism and an adequate decomposition of l are used. If $l \leq p \cdot r$ the calculation $\alpha^l \pmod{p(\alpha)}$ is made directly since it is computationally cheap. Else, l is rewritten as $l = l_2 + l_1(p^m)$ with l_1 as big as possible, as long as $l_1 < p * r$ and $m \leq \left\lfloor \frac{logl}{logp} \right\rfloor$, and therefore $\alpha^l = \alpha^{l_2}\left(\alpha^{l_1}\right)^{p^m}$. Thus, it is possible to compute α^{l_1} as a low-cost operation and use the Frobenius automorphism to calculate $\alpha^{l_1 \cdot p^m}$. Finally, we multiply times α^{l_2} and repeat the process many times as needed.

3.3 *Algorithm 3: Identifying Proteins as Codewords of BCH Codes*

Algorithm 3 is the final step in the process of identifying proteins as codewords of BCH codes. It will allow us to compute the generator polynomial $g(x)$ of the code C_{20} in such a way that, given a target sequence, it gets the cyclic codes with the largest minimum distance according to the BCH bound (Fig. 7).

To understand the important role of Algorithm 3, let us make a description of the protein identification process. Recall that, given a polypeptide sequence s_{20} that is identified as polynomial $s(x)$ in $\mathbb{Z}_{20}[x]$, we look for a generator polynomial $g_4(x)$ of a BCH code C_4 over $\mathbb{Z}_4$ and a generator polynomial $g_5(x)$ of a BCH code C_5 over $\mathbb{Z}_5$, such that $s_4(x)$ and $s_5(x)$ belong to the corresponding codes; and compute a polynomial $g(x)$ which is a generator of a cyclic code over $\mathbb{Z}_{20}$ that contains the

Algorithm 3: Calculate $g(x)$

On input: F_n, CL, $s(x)$ and n.
1: Find the minimal polynomials that divide $s(x)$: $P = \{f_i(x) \in F_n : f_i(x)|s(x), i \geq 0\}$
2: **For** $\{j \in \mathbb{N} : n > j > 0, gcd(j, n) = 1\}$
3: Apply BCH criterion: $(d, Fr_n) = BCH(P, CL, j)$, where Fr_n is the smallest set of minimal polynomials of P required to reach a minimal distance d.
4: **If** d is the biggest estimation of $d \geq 3$ **then**
5: Save Fr_n and d.
6: $g(x) = \prod_{f(x) \in Fr_n} f(x)$
7: **End if**
8: **End for**
9: **Return:** $Fr_n, F_n, g(x), d$

Fig. 7 Algorithm 3: Identifying proteins as codewords of BCH codes

polynomial $s(x)$, identified as a class in the polynomial quotient ring $\mathbb{Z}_{20}[x]/(x^n - 1)$.

To construct the polynomial $g_5(x)$, we apply Algorithm 1 ($p = 5$) to factor $x^n - 1$ and obtain the minimal polynomials. Then we apply Algorithm 3 to have a generator polynomial for the code C_5. On the other side, the construction of $s_4(x)$ is as follows. First, we apply Algorithm 1 for $p = 2$ to obtain the minimal polynomials, then we apply Graeffe's method to get the minimal polynomials over $\mathbb{Z}_4$, and finally we apply Algorithm 3 to have a generator polynomial for the code C_4.

For this last stage, the application of Algorithm 3 on the local ring $\mathbb{Z}_4$, the execution of step 3 of the algorithm is feasible thanks to important results such as Theorem 2.5, Theorem 3.4, and Theorem 3.5 of [1], or more generally, Corollary 16 in [7], which address the polynomial factorization, construction of minimal polynomials, and computation of the BCH bound for codes over finite commutative rings [1] and over chains of commutative rings [7]. Another work we can refer to is [8], in which through Theorem 4, the BCH bound for cyclic codes over rings of the form Z_{p^k} is established. Additionally, both [8, p. 482] and [7, p. 5] present algorithms for the construction of BCH codes given a designed distance. For its part, Algorithm 3 in [4] presents a strategy for finding the code with the largest minimum distance (using BCH bound) that identifies a given sequence $s(x)$. This involves an exhaustive search of all minimal polynomials. From the above, it is clear that Algorithm 3 can be applied to any finite local ring.

Another important aspect to consider is that Algorithm 3 constructs the generator polynomial $g(x)$ using some or all of the polynomials that divide $s(x)$, namely, those which belong to the set P constructed in step 1, to guarantee $s(x)$ belongs to the code. Finally, Algorithm 3 calculates the best possible BCH bound, rectifying the arbitrary election of the primitive polynomial in Algorithm 1. This choice does not have any impact on the generator polynomial $g(x)$ (nor on the code generated by $g(x)$), but it does on the BCH bound.

The following proposition is a construction that arose during the development of this work and is presented with a proof that came up naturally.

Proposition *Let α and β be primitive n-th roots of unity. $\mathfrak{M}_\alpha = \mathfrak{M}_\beta$, where $\mathfrak{M}_\alpha$ is the set of minimal polynomials of the factorization of $x^n - 1$ over $\mathbb{F}_q$ constructed using the primitive element α.*

Proof Consider a primitive element β. Since β is primitive, there exists $j \in \mathbb{N}$ such that $\mathrm{mcm}(j, n) = 1$ and $\beta^j = \alpha$. We introduce the following notation $f_i^\alpha(x)$ to denote the minimal polynomial in $\mathfrak{M}_\alpha$ associated with the q-cyclotomic coset of i. Then

$$\begin{aligned} f_i^\alpha(x) = \prod_{i\in\mathfrak{C}_i}\left(x-\alpha^i\right) &= \prod_{i\in\mathfrak{C}_i}\left(x-(\beta^j)^i\right) \\ = \prod_{\substack{z=ij\\ i\in\mathfrak{C}_i}}\left(x-\beta^z\right) &= \prod_{z\in j\mathfrak{C}_i}\left(x-\beta^z\right). \end{aligned} \tag{4}$$

Note that $j\mathfrak{C}_i = \mathfrak{C}_{ji}$. Therefore,

$$\prod_{z \in j\mathfrak{C}_i} (x - \beta^z) = \prod_{z \in \mathfrak{C}_{ij}} (x - \beta^z) = f_{ij}^{\beta}(x).$$

Which is to say, $\mathfrak{M}_\alpha \subseteq \mathfrak{M}_\beta$. The proof can be completed observing that $\mathfrak{M}_\beta \subseteq \mathfrak{M}_\alpha$ following the same steps and using the relation $\beta = \alpha^{j^{-1}}$. □

The aforementioned proposition shows that the same group of minimal polynomials is obtained no matter which primitive polynomial is chosen and that the q-cyclotomic coset to which a minimal polynomial is associated depends on the primitive element chosen for the construction of $\mathbf{H}_n$. The effect of considering a different primitive element β, where $\beta^j = \alpha$, is a configuration in which the minimal polynomial f_i^α is associated with the q-cyclotomic coset of $(j \cdot i) \mod n$. This is precisely the reason why the function in Algorithm 3 calculates the BCH bound for all possible configurations of minimal polynomials and q-cyclotomic cosets, ensuring that the best BCH bound is found.

4 Conclusions

The theoretical foundation for all three algorithms of [4] was given. Additionally, Proposition BDF was stated, which is a new result, as an improvement for Algorithm 1, reducing the number of required calculations.

Among the possibilities, for future work, there is the determination of exactly how much improvement Proposition BDF provides. Also, it might be interesting to consider different bounds such as the Hartmann-Tzeng (HT) bound for Algorithm 3.

In one of the steps of the algorithms presented in [8, p. 482] and [7, p. 5] for the construction of BCH codes, the minimal polynomials are considered. Therefore, it would be interesting to review the contribution of the Proposition BDF regarding the efficiency in the execution of this step.

References

1. Andrade, A., Palazzo, R.: Construction and decoding of BCH codes over finite commutative rings. Linear Algebra Appl. **286**(1), 69–85 (1999)
2. Duarte, M.E., Echeverri, O., Guevara, J.M., Palazzo, R.: Cyclic concatenated genetic encoder: A mathematical proposal for biological inferences. Biosystems **163**, 12 (2017)
3. Fu, Q., Li, R.H., Guo, L.B.: LCD MDS codes from cyclic codes. Procedia Comput. Sci. **154**, 663–668 (2019). Proceedings of the 9th International Conference of Information and Communication Technology [ICICT-2019] Nanning, Guangxi, China January 11–13, 2019

4. Galíndez, V., Duarte, M.E.: Construction of Cyclic Codes over $\mathbb{Z}_{20}$ for Identifying Proteins. Applied Computer Sciences in Engineering. WEA 2019. Communications in Computer and Information Science **1052**, pp. 40–51 (2019)
5. Huffman, W.C., Pless, V.: Fundamentals of Error-Correcting Codes. Cambridge University Press, Cambridge (2003)
6. McDonald, B.R.: Finite rings with identity. M. Dekker, New York (1974)
7. Shah, T., Qamar, A., Andrade, A.: Construction and decoding of BCH codes over chain of commutative rings. Math. Sci. **6**(51), 1–14 (2012)
8. Shankar, P.: On BCH codes over arbitrary integer rings (corresp.). IEEE Trans. Inf. Theory **25**(4), 480–483 (1979)
9. Solé, P.: A quaternary cyclic code, and a family of quadriphase sequences with low correlation properties. Coding Theory and Applications. Coding Theory 1988 **388**, pp. 193–201 (1988)
10. Uspensky, J.V.: Teoría de Ecuaciones. Editorial Limusa S.A. (1995)
11. Wesley Peterson, W., Weldon, E.J.: Error-correcting Codes, 2nd edn. MIT Press Cambridge (1972)

Uniqueness of the Star Central Configuration in the 5-Body Problem

Leasly A. Campa-Raymundo

Abstract In this note, we study central configurations in the 5-body problem with equal masses. We aim to show that the only central configuration $\mathbf{q} = (\mathbf{q}_1, \ldots, \mathbf{q}_5)$ satisfies that each $\mathbf{q}_i$ is placed at a vertex of a polygon such that the central angle is $\frac{2\pi}{5}$ and $q_1 = (1, 0)$. This configuration corresponds to a regular polygon.

Keywords Central configuration · Regular polygon configuration · 5-Body problem

1 Introduction

The n-body problem is central to celestial mechanics as it involves studying the motion of n point masses interacting solely through universal gravitational forces. The case $n = 2$ has been solved, with the two bodies following conic trajectories, but solutions for $n \geq 3$ are limited to specific scenarios, known as homographic solutions, which remain invariant under rotations and scaling. Moreover, they are the only known explicit solutions for the n-body problem. Therefore, central configurations are counted up to these motions [1].

This paper is structured as follows: in Sects. 2 and 3, we cover basic concepts and results related to central configurations and introduce some helpful ones for the rest of the paper. Section 4 focuses on the aim of this paper, which is the existence and uniqueness of the solution of the star central configuration (SCC) (6). To achieve this, we first establish the existence of the solution of the SCC in Theorem 1. After that, in Sect. 3.1 we simplify the problem to two variables, resulting in a new significant domain $\mathcal{J}$ in the plane. To prove the uniqueness of the SCC, in Sect. 4 we divide the domain $\mathcal{J}$ into 16 disjoint regions for which certain equations (see (7)) are not satisfied in each region.

L. A. Campa-Raymundo (✉)
Departamento de Matemáticas, Universidad Autónoma Metropolitana, Mexico City, Mexico
e-mail: lcampar@xanum.uam.mx

P. E. Harris et al. (eds.), *Women in Mathematics in Latin America*,
Research Perspectives Ghent Analysis and PDE Center 14,
https://doi.org/10.1007/978-3-031-97857-9_5

2 Preliminaries

The n-body problem determines the motion of n particles without volume in $\mathbb{R}^d$ ($d = 1, 2, 3$), as they move under the influence of Newton's law of gravitation. By assuming the gravitational constant $G = 1$ having particles with positive masses m_i and positions $\mathbf{q}_i \in \mathbb{R}^d$, the equations of motion can be expressed as follows:

$$m_i \ddot{\mathbf{q}}_i = - \sum_{i=1, i \neq j}^{n} \frac{m_i m_j}{r_{ij}^3} (\mathbf{q}_i - \mathbf{q}_j) = \frac{\partial U}{\partial \mathbf{q}_i}, \quad i = 1, 2, \ldots, n, \tag{1}$$

being $\ddot{q}$ the second derivative of q_i with respect to time. The Newtonian potential is the function $U : X \to \mathbb{R}$, given by

$$U = \sum_{i<j} \frac{m_i m_j}{r_{ij}},$$

where $r_{ij} = |\mathbf{q}_i - \mathbf{q}_j|$ is the mutual distance in the Euclidean norm and the position vector is $\mathbf{q} = (\mathbf{q}_1, \ldots, \mathbf{q}_n) \in \mathbb{R}^{nd}$. Let us define the sets $\Delta_{ij} = \{\mathbf{q} \in \mathbb{R}^{dn} \colon \mathbf{q}_i = \mathbf{q}_j, i \neq j\}$ which indicates all binary collisions, and the collision set by $\Delta = \cup_{i \neq j} \Delta_{ij}$. Thus, the configuration space corresponds to $X = \mathbb{R}^{dn} \setminus \Delta$.

In what follows, we assume that the center of mass is located at the origin, that is $m_1 \mathbf{q}_1 + \ldots + m_n \mathbf{q}_n = \mathbf{0}$, since it is the first integral for the system.

A *central configuration (CC)* in the n-body problem for given masses and some fixed time t_0, satisfies for $k = 1, \ldots, n$ that

$$-\ddot{\mathbf{q}}_k(t_0) = \lambda \mathbf{q}_k(t_0), \quad \lambda \in \mathbb{R}. \tag{2}$$

A CC $\mathbf{q} = (\mathbf{q}_1, \ldots, \mathbf{q}_n)$ is called **Star Central Configuration (SCC)** if each $\mathbf{q}_i$, with $i = 1, \ldots, n$, is placed at a vertex of a polygon for which the central angle θ (the angle between two consecutive ratios) is $\frac{2\pi}{n}$.

The SCC are easy to identify in polar coordinates because they can be written as

$$\mathbf{q}_i = r_i(\cos(\theta_i), for \sin(\theta_i)) = (q_{i1}, q_{i2}), \tag{3}$$

where $\theta_i = \frac{2\pi(i-1)}{n}$, for $i = 1, \ldots n$.

If the configuration $\mathbf{q}_0$ is a CC, then so are $c\mathbf{q}_0$ and $A\mathbf{q}_0$ for any $c \in \mathbb{R}^+$ and $A \in SO(3)$. In this way, homothecies and rotations of a Central Configuration will also be Central Configuration. This is a natural way to count for the CCs.

Consider $M\ddot{\mathbf{q}}_0 = \nabla U(\mathbf{q}_0)$, where M is the mass matrix $M = \text{diag}\{m_1, \ldots, m_n\}$. By applying the matrix M in both sides of this definition and using the equations of motions (1), we obtain $\nabla U(\mathbf{q}_0) = \lambda \nabla I(\mathbf{q}_0)$, which induces a multiplier Lagrange problem, being λ the Lagrange multiplier. In this way, the CCs satisfy a maximum (optimization) problem of the function U restricted to I.

Since $c\mathbf{q}_0$ and $A\mathbf{q}_0$ are CC, we can count classes of CC by using a equivalent relation. Therefore, we normalize CC by setting $I = 1$. So, the configuration $\mathbf{q}_0$ is a normalized CC if and only if it is a critical point of the restriction of $U(\mathbf{q})$ to the sphere of masses (given by $I = 1$). It is also important to note that the restriction of the potential U to S always attains its minimum at some $\mathbf{q}_0 \in S$.

The function IU^2 is called *configuration measure*, with two features: it is homogeneous of degree zero, and the functions U and I depend only on the mutual distances. Thus, they are also invariant under rotations. Therefore, IU^2 measures how the shape of the configuration is changing. Another characterization is that the configuration $\mathbf{q}_0$ is CC if only if it is a critical point of IU^2.

3 Basic Results

Let us consider

$$\mathcal{S} = \{\mathbf{q} \in X : I = 1\}\,, \tag{4}$$

which is known as the *sphere of masses*, where I is the moment of inertia,

$$I(\mathbf{q}) = \frac{1}{2}\sum_{j=1}^{n} m_j \mathbf{q}_j^2 = \frac{1}{4\tilde{m}}\sum_{i<j} m_i m_j r_{ij}^2,$$

with total mass $\tilde{m} = m_1 + \ldots + m_n$.

Also, we consider the following set:

$$\hat{\mathcal{J}} = \{\mathbf{q} \in X : \mathbf{q}_1 = (1, 0)\}\,. \tag{5}$$

It is well-known that a critical value of U is a critical value of IU^2 [2]. As for the sphere of masses the point $|\mathbf{q}_1|^{-1}\mathbf{q}$ is critical for $U|_{\hat{\mathcal{J}}}$ if and only if $\mathbf{q} \in \mathcal{S}$ is a CC. Hence, the restriction of the potential U to $\hat{\mathcal{J}}$ always attains its minimum at some $\mathbf{q} \in \hat{\mathcal{J}}$.

3.1 5-Body Problem

We regard 5 bodies with equal masses, which conform a SCC in $\hat{\mathcal{J}}$,

$$\lambda \mathbf{q}_i = m\sum_{i\neq j}^{5} \frac{\mathbf{q}_i - \mathbf{q}_j}{r_{ij}^3}. \tag{6}$$

Taking into account polar coordinates given in (3), and by using the center of mass fixed at the origin, we can reduce (6) to the equations

$$\lambda_{ik}(r_3, r_5) = \frac{m}{q_{ik}} \sum_{i \neq j} \frac{q_{ik}(r_3, r_5) - q_{jk}(r_3, r_5)}{r_{ij}^3}, \quad k = 1, 2. \tag{7}$$

$$\lambda_{12}(r_3, r_5) = 0, \quad \text{with} \quad q_{ik} \neq 0,$$

whence the parameter λ is expressed as functions depending on r_3 and r_5. According to the definition of CC, $\lambda_{i,k} = \lambda_{l,m}$ for all $i, k, l, m \in \{1, 2\}$. So, the domain $\hat{\mathcal{J}}$ for the system (7) is described by the set

$$\hat{\mathcal{J}} = \left\{ (r_3, r_5) \in \mathbb{R}^2 \colon r_3 > 0, r_5 > 0, r_5 > r_3 - b/2, r_5 > \frac{ar_3 - a}{2} \right\}, \tag{8}$$

where $a = \sqrt{5} + 1$ and $b = \sqrt{5} - 1$.

We are ready to tackle the main goal of this work, which is to show the existence and uniqueness of the solution of SCC (6). The above reasoning reduces this problem to (7).

Remark 1 It can be readily verified that the regular pentagon corresponds to an SCC that satisfies (7).

The last remark asserts the existence of a solution of (7). It remains to show the uniqueness of the solution, which will be addressed in the following section.

4 Uniqueness

It is important to emphasize that the point $(r_3, r_5) = (1, 1)$ is a minimum value of the function IU^2, which implies the existence of the solution (see Remark 1). The following theorem introduces the main result of this paper.

Theorem 1 *There is no solutions to the system* (7), *for* $(r_3, r_5) \neq (1, 1)$.

The proof of Theorem 1 requires dividing the domain $\hat{\mathcal{J}}$ into 16 regions (see Fig. 1), in which the CC equations are not satisfied. Namely, for each $\mathcal{J}_n$, with $n = 1, \ldots, 16$, there exist $\lambda_{ij}(r_3, r_5)$ and $\lambda_{kl}(r_3, r_5)$ given in (7), for which $\lambda_{ij}(r_3, r_5) \neq \lambda_{kl}(r_3, r_5)$, for all r_3 and r_5 in $\mathcal{J}_n$. In the following, we make an analytical reasoning of these regions.

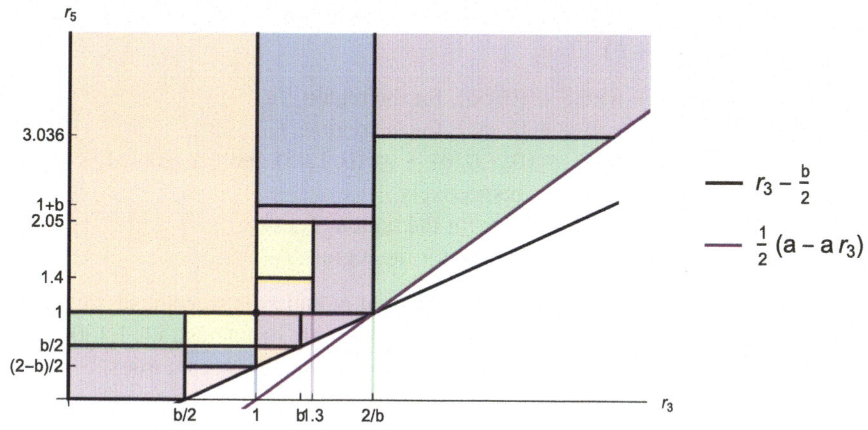

Fig. 1 Regions $\mathcal{J}_1, \ldots, \mathcal{J}_{16}$

4.1 Analytical Reasoning of the Regions $\mathcal{J}_n$

Taking into account the equations of system (7), we consider families of functions such that any two different families do not intersect in any region.

For the regions $\mathcal{J}_1$ and $\mathcal{J}_2$, we define the family of functions $\lambda_{ik_\eta}(r_5) :- \lambda_{ik}\left(\frac{b}{2+\eta}, r_5\right)$, with $\eta \in [0, \infty)$. The following statements are true:

- $\lambda_{11_\eta}(r_5) < \lambda_{31_\eta}(r_5)$, for the region $\mathcal{J}_1$.
- $\lambda_{41_\eta}(r_5) < \lambda_{31_\eta}(r_5)$, for the region $\mathcal{J}_2$.

For the regions $\mathcal{J}_3$, $\mathcal{J}_4$ and $\mathcal{J}_5$ if we define the following family of functions $\lambda_{ik_\zeta}(r_3) :- \lambda_{ik}\left(r_3, \frac{b}{2+\zeta}\right)$, then following assertions are true:

- $\lambda_{11_\zeta}(r_3) < \lambda_{52_\eta}(r_3)$, with $\zeta \in [0, nz)$, for the region $\mathcal{J}_3$.
- $\lambda_{11_\zeta}(r_3) < \lambda_{52_\eta}(r_3)$, with $\zeta \in [nz, nd(\zeta))$, for the region $\mathcal{J}_4$.

For the region $\mathcal{J}_6$, we defined the following family of functions $\lambda_{ik_\mu}(r_5) :- \lambda_{ik}(b + \mu, r_5)$, with $\mu \in \left[0, \frac{2}{b} - b\right)$, and we obtain the following:

- $\lambda_{11_\mu}(r_5) > \lambda_{31_\eta}(r_5)$.

For the regions $\mathcal{J}_7$ and $\mathcal{J}_8$, we consider the following family of functions $\lambda_{ik_\iota}(r_3) :- \lambda_{ik}(r_3, 1 - \iota)$, with $\iota \in (0, 1 - b/2)$, and the following statements are true:

- $\lambda_{52_\iota}(r_3) > \lambda_{42_\iota}(r_3)$, for the region $\mathcal{J}_7$, while for $\mathcal{J}_8$
- The family $\lambda_{22_\iota}(r_3) - \lambda_{32_\iota}(r_3)$ is strictly positive.

For the regions $\mathcal{J}_9$ to $\mathcal{J}_{15}$, we defined the following family of functions $\lambda_{ik_\xi}(r_3) :- \lambda_{ik}(r_3, 1+\xi)$. Thus,

- $\lambda_{11_\xi}(r_3) > \lambda_{41_\xi}(r_3)$, with $\xi \in (0, \infty)$, for the region $\mathcal{J}_9$.
- $\lambda_{11_\xi}(r_3) > \lambda_{41_\xi}(r_3)$, with $\xi \in [b, \infty)$, for the region $\mathcal{J}_{10}$.
- $\lambda_{11_\xi}(r_3) > \lambda_{51_\xi}(r_3)$ for $\xi \in (0, 2.036)$, $\xi \in (0, 1.05)$ and $\xi \in (0.4, 1.05)$ into the regions $\mathcal{J}_{11}$, $\mathcal{J}_{12}$ and $\mathcal{J}_{13}$, respectively.
- $\lambda_{11_\xi}(r_3) > \lambda_{31_\xi}(r_3)$, $\xi \in [1.05, b)$, for the region $\mathcal{J}_{14}$.
- $\lambda_{11_\xi}(r_3) > \lambda_{52_\xi}(r_3)$, $\xi \in [2.036, \infty)$, for the region $\mathcal{J}_{15}$.

To conclude, for the region $\mathcal{J}_{16}$ the domain of r_3 and r_5 is strategically divided, in such a way that only the following intervals of r_3 are of interest: $(1, 1.13067] \cup (1.13067, 1.52781) \cup (1.201923, 1.3]$. The following statements are true:

- For $r_3 \in [1, 1.13067]$ and $r_5 \in (1, 1.4)$, we have $\lambda_{21_{r_3}}(r_5) > \lambda_{41_{r_3}}(r_5)$.
- For $r_3 \in (1.13067, 1.152781)$ and $r_5 \in (1, 1.4)$, we have $\lambda_{11_{r_3}}(r_5) > \lambda_{21_{r_3}}(r_5)$.
- For $r_3 \in (1.201923, 1.3)$ and $r_5 \in (1, 1.4)$, we have $\lambda_{11_{r_3}}(r_5) > \lambda_{31_{r_3}}(r_5)$.

It is worth mentioning that we omitted some proofs and rigorous calculations, as they will appear in a forthcoming manuscript.

References

1. Saari, D.G.: Collisions, rings, and other Newtonian N-body problems. In: CBMS Regional Conference Series in Mathematics, vol. 104 (American Mathematical Society, Providence, 2005)
2. Moeckel, R.: Central configurations. In: Central Configurations, Periodic Orbits, and Hamiltonian Systems, Adv. Courses Math. CRM Barcelona, pp. 105–167. Birkhäuser/Springer, Basel (2015)

Uniform Temperature Distribution Induced by an Optimal Domain Shape

Andrea Ceretani, Cuiyu He, Weiwei Hu, Lin Mu, and Carlos N. Rautenberg

Abstract This chapter is devoted to present some recent results in shape optimization problems involving heat-conductive flows. We focus on the optimal shape design of a two dimensional container to achieve a temperature distribution as uniform as possible. Our geometrical setting allows to prove the well-posedness to the Boussinesq system, the constraint, for each admissible domain. Moreover, the weak solution has $H^{3/2}$ regularity on the boundary portion where (smooth) deformations are allowed. Further, we prove the existence of a solution to the shape optimization problem of interest. We finally present a directional differentiability result and deduce a first-order optimality condition.

Keywords Shape optimization · Heat-conductive flows · Navier-Stokes equations

---

A. Ceretani (✉)
Departamento de Matemática, Facultad de Ciencias Exactas y Naturales, Universidad de Buenos Aires (UBA), and Instituto de Investigaciones Matemáticas Luis A. Santaló (IMAS), UBA-CONICET, Buenos Aires, Argentina
e-mail: aceretani@dm.uba.ar

C. He · W. Hu · L. Mu
Department of Mathematics, University of Georgia, Athens, GA, USA
e-mail: cuiyu.he@uga.edu; weiwei.hu@uga.edu; lin.mu@uga.edu

C. N. Rautenberg
Department of Mathematical Sciences and the Center for Mathematics and Artificial Intelligence (CMAI), George Mason University, Fairfax, VA, USA
e-mail: crautenb@gmu.edu

P. E. Harris et al. (eds.), *Women in Mathematics in Latin America*,
Research Perspectives Ghent Analysis and PDE Center 14,
https://doi.org/10.1007/978-3-031-97857-9_6

1 Introduction

The problem of interest consists of determining the optimal shape for a two-dimensional region containing a heat-conductive incompressible fluid, and with the goal to achieve a uniform temperature distribution. More precisely, we consider domains of the form

$$\Omega_\gamma = \{(x_1, x_2) \in \mathbb{R}^2 \,:\, 0 < x_1 < 1,\ \gamma(x_1) < x_2 < 1\}, \tag{1}$$

and look for an optimal choice of the function γ; see Fig. 1.

We assume that the domain has constant temperature equal to zero on top and lateral walls, and a warmed bottom boundary. Notice that the heated part of the domain boundary is precisely the place where shape perturbations are allowed. Also, we assume that the fluid does not slip at the fluid-solid interface. Then the flow is generated by buoyancy due to temperature changes, which we model by the Boussinesq approximation. Thus, the velocity field $\boldsymbol{v}$ and temperature distribution T of the fluid are governed by

$$\boldsymbol{v} \cdot \nabla \boldsymbol{v} - \frac{1}{\mathrm{Re}} \Delta \boldsymbol{v} + \nabla p = \frac{\mathrm{Gr}}{\mathrm{Re}^2} T \boldsymbol{e} \quad \text{in} \quad \Omega_\gamma, \tag{2}$$

$$\nabla \cdot \boldsymbol{v} = 0 \quad \text{in} \quad \Omega_\gamma, \tag{3}$$

$$\boldsymbol{v} = 0 \quad \text{on} \quad \partial\Omega_\gamma, \tag{4}$$

and

$$\boldsymbol{v} \cdot \nabla T - \frac{1}{\mathrm{RePr}} \Delta T = 0 \quad \text{in} \quad \Omega_\gamma, \tag{5}$$

$$T = T_d \quad \text{on} \quad \Gamma_\gamma, \tag{6}$$

$$T = 0 \quad \text{on} \quad \partial\Omega_\gamma \setminus \Gamma_\gamma. \tag{7}$$

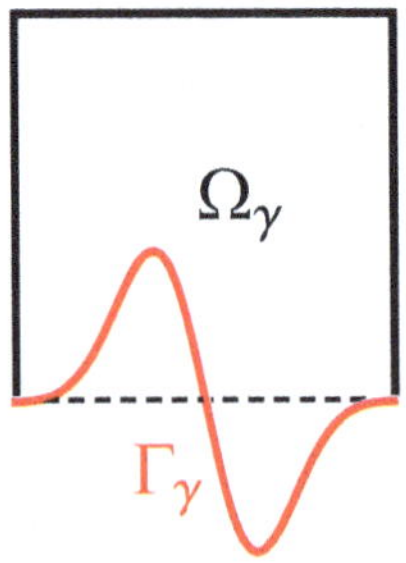

Fig. 1 Sketch of a possible admissible domain

Here, $\partial\Omega_\gamma$ indicates the boundary of Ω_γ and Γ_γ represents its bottom part. Specifically,

$$\Gamma_\gamma = \{(x_1, x_2) \in \partial\Omega_\gamma \, : \, x_2 = \gamma(x_1)\}. \tag{8}$$

Also, $\mathbf{e} = (0, 1)$, T_d is the prescribed temperature on Γ_γ, and Re, Gr, Pr are the Reynolds, Grashof and Prandtl numbers. The ambient temperature is assumed to be zero.

The shape optimization problem of interest is given by

$$\begin{aligned} &\text{Find } \gamma^* \in V \text{ such that } J(T^*, \gamma^*) = \inf_{\gamma \in V} J(T, \gamma), \\ &\text{subject to } (\boldsymbol{v}, T) \text{ is the weak solution to (2)–(7) in } \Omega_\gamma, \end{aligned} \tag{9}$$

where

$$J(T, \gamma) = \|T - I(T, \gamma)\|^2_{L^2(\Omega_\gamma)} + \frac{\lambda}{2}\|\gamma'\|^2_{L^2(0,1)}, \tag{10}$$

for some given $\lambda > 0$, and

$$I(T, \gamma) = \int_{\Omega_\gamma} T(x)\, dx. \tag{11}$$

Also, V is the set of all functions γ that satisfy

1. $\gamma \in H^3(0, 1)$,
2. $\gamma^{(\alpha)}(x) = 0$ for $\alpha = 0, 1, 2$, and all $x \in (0, \varepsilon) \cup (1 - \varepsilon, 1)$,
3. $\gamma \leq \nu$ a.e. in $(0, 1)$ and $|\gamma'| \leq L$ a.e. in $(0, 1)$,
4. $\int_0^1 \gamma(\xi)\, d\xi = 0$,

for some given $L > 0$ and $\varepsilon, \nu \in (0, 1)$.

The definition of the cost function J given in (10) is related to minimizing the variance of the temperature (first term) as well as to prevent strong oscillations of the function γ (second term).

A few words are in order concerning the choice of the admissible set V for the function γ. First, observe that Condition 2 involving pointwise evaluations of second order derivatives is proper given that $\gamma \in H^3(0, 1) \hookrightarrow C^2_B(0, 1)$. Further, Conditions 1 and 2 together, imply that Γ_γ is of class C^2. From this, we are able to derive boundary regularity properties of the weak solution to the Boussinesq system which are required to compute the shape derivative of the objective functional. Condition 3 is related to the existence of an optimal domain. In particular, this condition implies that $\Omega^{\mathrm{m}} \subset \Omega_\gamma \subset \Omega^{\mathrm{M}}$ where $\Omega^{\mathrm{m}} = (0, 1) \times (\nu, 1)$ and $\Omega^{\mathrm{M}} = (0, 1) \times (-M, 1)$ for some $M > 0$, from which we shall obtain domain estimates for the weak solution to the Boussinesq system that are independent of γ. Finally, Condition 4 restricts the set of admissible domains to those having a unitary measure. Then the average temperature inside the domain is given by $I(T, \gamma)$. The absence of the factor $|\Omega_\gamma|^{-1}$ in the definition of the average temperature significantly simplifies the computation of the directional derivative of the reduced objective functional $\hat{J}(\gamma) := J(T(\gamma), \gamma)$, where $T(\gamma)$ is the weak temperature in Ω_γ.

The outline of the chapter is as follows. In Sect. 2, we present an existence result for the solution to the shape optimization problem (9) (see Theorem 2). The well-posedness of the Boussinesq system (2)–(7) and some boundary regularity properties of its weak solution are also given in this section (see Theorem 1). Then, we present a first-order optimality condition in Sect. 3 (see Theorem 3). There, we further present the adjoint system to (2)–(7) and its well-posedness (see Lemma 1). Finally, we discuss our ongoing and future work in Sect. 4.

The results of this chapter extend naturally to a more general setting where external forces or heat sources are allowed within the domain. In addition, several results presented here still hold true under milder assumptions but stronger ones are considered to present our results in a unified framework, where all theorems hold true.

Optimization problems involving the Navier-Stokes equations as constraints are usually challenging. The nonlinear nature of the Navier-Stokes equations along the (in general) low regularity of the spatial domain, the possible coupling with transport equations, or the presence of mixed boundary conditions, among others, affect the mathematical analysis of them. Often, all these aspects must be handle with care to achieve good regularity properties for the variables describing the fluid (e.g. velocity, pressure, temperature), which are required to address the optimization problem of interest. Several contributions have been made in the last decades for problems constrained by the Stokes, Navier-Stokes or Boussinesq equations, see for example [6, 7, 9, 10, 16, 18–20, 23] and the references therein.

2 The Boussinesq System and the Shape Optimization Problem

In this section, we focus on the well-posedness of the Boussinesq system (2)–(7) and the existence of solution to the shape optimization problem (9). The existence, uniqueness, and regularity properties of a solution to (2)–(7) are given in Theorem 1. The existence of solution to (9) is presented in Theorem 2. We do not present detailed proofs for these theorems. However, after each of them, we discuss the main proof arguments.

We begin by introducing the weak formulation of (2)–(7). From now on we assume that the boundary data T_d is the restriction in the trace sense of a function in $H^1(\Omega^{\mathrm{M}}) \cap C_c(\Omega^{\mathrm{M}})$. To maintain a clear notation, we denote the later by T_d as well. Notice that $T_d \in H^{1/2}(\partial\Omega_\gamma)$ and $T_d = 0$ on $\partial\Omega_\gamma \setminus \Gamma_\gamma$.

We say that $(\boldsymbol{v}, T) \in V(\Omega_\gamma) \times H^1(\Omega_\gamma)$ is a weak solution to (2)–(7) if $T - T_d \in H_0^1(\Omega_\gamma)$ and

$$\int_{\Omega_\gamma} (\boldsymbol{v} \cdot \nabla \boldsymbol{v}) \cdot \boldsymbol{\varphi}\, dx + \frac{1}{\mathrm{Re}} \int_{\Omega_\gamma} \nabla \boldsymbol{v} \cdot \nabla \boldsymbol{\varphi}\, dx = \frac{\mathrm{Gr}}{\mathrm{Re}^2} \int_{\Omega_\gamma} T\mathbf{e} \cdot \boldsymbol{\varphi}\, dx \quad \forall \boldsymbol{\varphi} \in V(\Omega_\gamma), \tag{12}$$

$$\int_{\Omega_\gamma} (\boldsymbol{v} \cdot \nabla T)\varphi\, dx + \frac{1}{\mathrm{RePr}} \int_{\Omega_\gamma} \nabla T \cdot \nabla \varphi\, dx = 0 \qquad \forall \varphi \in H_0^1(\Omega_\gamma), \tag{13}$$

where

$$V(\Omega_\gamma) = \{\boldsymbol{\varphi} \in H^1(\Omega_\gamma) : \operatorname{div} \boldsymbol{\varphi} = 0 \text{ a.e. in } \Omega_\gamma, \boldsymbol{\varphi} = 0 \text{ on } \partial\Omega_\gamma \text{ in the trace sense}\},$$
$$H_0^1(\Omega_\gamma) = \{\varphi \in H^1(\Omega_\gamma) : \varphi = 0 \text{ on } \partial\Omega_\gamma \text{ in the trace sense}\}.$$

The function spaces $V(\Omega_\gamma)$ and $H_0^1(\Omega_\gamma)$ are equipped here with the norms

$$\|\boldsymbol{\varphi}\|_{V(\Omega)} = \left(\int_\Omega |\nabla \boldsymbol{\varphi}|^2\, dx\right)^{1/2} \quad \text{and} \quad \|\varphi\|_{H_0^1(\Omega)} = \left(\int_\Omega |\nabla \varphi|^2\, dx\right)^{1/2},$$

respectively.

We are now in a position to address the well-posedness of the Boussinesq system in the next theorem. As usual, it requires moderate values for the Reynolds, Prandtl, and Grashof numbers, and a smallness condition on the boundary data.

Theorem 1 (Weak Solution to the Boussinesq System) *Let $T_d \in H^1(\Omega^{\mathrm{M}}) \cap C_c(\Omega^{\mathrm{M}})$ and $\gamma \in V$. If $T_d \in H^{3/2}(\partial\Omega_\gamma)$, then there exist $\varepsilon_1, \varepsilon_2, \varepsilon_3 > 0$ such that if*

$$\mathrm{Re} \in (0, \varepsilon_1), \qquad \mathrm{Pr} \in \left(0, \frac{\varepsilon_2}{\mathrm{Re}}\right), \qquad \mathrm{Gr} \in (0, \varepsilon_3 \mathrm{Re}), \tag{14}$$

and T_d is sufficiently small, then (2)–(7) *admits a unique weak solution $(\boldsymbol{v}, T) \in V(\Omega_\gamma) \times H^1(\Omega_\gamma)$, which satisfies*

$$\|\boldsymbol{v}\|_{V(\Omega_\gamma)} + \|T - T_d\|_{H_0^1(\Omega_\gamma)} \le \tau, \tag{15}$$

where τ is independent of γ. In addition, $\boldsymbol{v}|_{\Gamma_\gamma^\varepsilon} \in H^{3/2}(\Gamma_\gamma^\varepsilon)$ and $T|_{\Gamma_\gamma^\varepsilon} \in H^{3/2}(\Gamma_\gamma^\varepsilon)$ for any $\varepsilon \in (0, 1)$, where $\Gamma_\gamma^\varepsilon = \{(x_1, x_2) \in \Gamma_\gamma : \varepsilon < x_1 < 1 - \varepsilon\}$.

The existence and uniqueness of a weak solution to (2)–(7) is obtained from a fixed point argument, similarly as in [2]. The smallness condition on T_d is in the $H^1(\Omega^{\mathrm{M}})$-sense and it is related to condition (14) for the Reynolds, Prandtl, and Grashof numbers. These conditions are required to prove that the fixed point map is a contraction from some closed ball into itself. Estimation (15) is then a direct consequence of the Banach's fixed point theorem. This part of the theorem still holds true if $T_d|_{\Gamma_\gamma^\varepsilon} \in H^{1/2}(\partial\Omega)$. However, the extra regularity required for T_d leads to the improved boundary regularity for the weak solution on any subset $\Gamma_\gamma^\varepsilon$. The proof for $\boldsymbol{v}$ follows from a bootstrap argument that exploits properties of a related Stokes problem, see [12, Thm. IV.5.1, p. 276], while the proof for T relies on well-known properties of the Poisson problem for the Laplacian with homogeneous Dirichlet conditions, see [11, Thm. 4, p. 334].

It is worth to mention that the parameter τ can be chosen to be independent of γ. The proof reduces to show that some constants, involved in the contractivity proof part regarding the fixed point mapping, can be set independently of γ, see [2]. These constants come from the Sobolev embedding and Poincaré inequality theorems in Ω_γ. In our setting, due to the inclusions $\Omega^{\mathrm{m}} \subset \Omega_\gamma \subset \Omega^{\mathrm{M}}$ for any $\gamma \in V$, that constants can be chosen so that they do not depend on Ω_γ, and therefore on γ; see [1, 14].

The next result concerns the existence of a solution to problem (9).

Theorem 2 (Existence of an Optimal Domain) *Let $T_d \in H^1(\Omega^{\mathrm{M}}) \cap C_c(\Omega^{\mathrm{M}})$ and $\gamma \in V$. If* Re, Pr, Gr *satisfy* (14) *and T_d satisfies the smallness condition in Theorem 1, then there exists a solution $\gamma^* \in V$ to the shape optimization problem* (9).

This theorem can be proven with the same spirit as in [9]. Given an infimizing sequence $\{\gamma_n\}$, we consider the associated sequence of weak solutions $\{(\boldsymbol{v}_n, T_n)\}$ to (2)–(7) in Ω_{γ_n}, where $T_n = \hat{T}_n + T_d$ and $\hat{T}_n \in H_0^1(\Omega_{\gamma_n})$. Then, we extend each pair $(\boldsymbol{v}_n, \hat{T}_n)$ by zero outside Ω_{γ_n} and deduce that the sequence of extensions converges weakly to some $(\boldsymbol{v}^{\mathrm{M}}, \hat{T}^{\mathrm{M}}) \in H^1(\Omega^{\mathrm{M}}) \times H^1(\Omega^{\mathrm{M}})$ along a subsequence. Here we use the estimate (15) and exploit that τ does not depend on the domain. Subsequently, we consider the restriction $(\boldsymbol{v}^*, \hat{T}^*) \in H^1(\Omega_{\gamma^*}) \times H^1(\Omega_{\gamma^*})$ of $(\boldsymbol{v}^{\mathrm{M}}, \hat{T}^{\mathrm{M}})$ to Ω_{γ^*} and prove that $(\boldsymbol{v}^*, T^*)$ is a weak solution to the Boussinesq system (2)–(7) in Ω_{γ^*}, where $T^* = \hat{T}^* + T_d$, $\hat{T}^* \in H_0^1(\Omega^*)$ and Ω_{γ^*} is the limit of $\{\Omega_{\gamma_n}\}$ in the Hausdorff complementary metric. Finally, we deduce the optimality of γ^* from the weak lower semicontinuity of the objective functional J.

3 The Adjoint System and a First-Order Optimality Condition

We now turn our attention to deriving a first-order optimality condition for the shape optimization problem (9). Hereafter, we assume that Re, Pr, Gr, and T_d are under the assumptions of Theorem 1.

The directional derivative of the reduced objective functional is deduced from the adjoint approach described in [17]. The next lemma provides the required preliminary framework. Namely, the computation and well-posedness of the adjoint system.

Lemma 1 *Let* $T_d \in H^1(\Omega^{\mathrm{M}}) \cap C_c(\Omega^{\mathrm{M}})$ *and* $\gamma \in V$. *If* Re, Pr, Gr *satisfy* (14) *and* T_d *satisfies the smallness condition in Theorem 1, then the adjoint system associated to* (2)–(7) *in* Ω_γ *is given by*

$$-\boldsymbol{v}\cdot\nabla\boldsymbol{w}+\boldsymbol{w}\cdot\nabla\boldsymbol{v}+S\,\nabla T-\frac{1}{\mathrm{Re}}\Delta\boldsymbol{w}+\nabla q=0 \quad \text{in} \quad \Omega_\gamma, \tag{16}$$

$$\nabla\cdot\boldsymbol{w}=0 \quad \text{in} \quad \Omega_\gamma, \tag{17}$$

$$\boldsymbol{w}=0 \quad \text{on} \quad \partial\Omega_\gamma, \tag{18}$$

and

$$-\boldsymbol{v}\cdot\nabla S-\frac{1}{\mathrm{RePr}}\Delta S-\frac{\mathrm{Gr}}{\mathrm{Re}^2}\boldsymbol{w}\cdot\mathbf{e}=T-I(T,\gamma) \quad \text{in} \quad \Omega_\gamma, \tag{19}$$

$$S=0 \quad \text{on} \quad \partial\Omega_\gamma, \tag{20}$$

where $(\boldsymbol{v}, T)$ *is the weak solution to* (2)–(7) *in* Ω_γ *given by Theorem 1. In addition, there exist* $\varepsilon_1, \varepsilon_2, \varepsilon_3 > 0$ *such that if*

$$\mathrm{Re}\in(0,\varepsilon_1), \qquad \mathrm{Pr}\in\left(0,\frac{\varepsilon_2}{\mathrm{Re}}\right), \qquad \mathrm{Gr}\in(0,\varepsilon_3\min(\mathrm{Re},\mathrm{Re}^2)), \tag{21}$$

and T_d *is sufficiently small, then there exists a unique weak solution* $(\boldsymbol{w}, S) \in V(\Omega_\gamma)\times H_0^1(\Omega_\gamma)$ *to* (16)–(20) *in* Ω_γ, *for which* $\boldsymbol{w} \in H^{3/2}(\Gamma_\gamma^\varepsilon)$ *and* $S \in H^{3/2}(\Gamma_\gamma^\varepsilon)$ *for any* $\varepsilon \in (0, 1)$.

We do not provide the details of the computations leading to (16)–(20) since they are quite technical and only involve standard arguments. The well-posedness of the adjoint system (16)–(20) is obtained, as before, from a fixed point argument. A key step of this part of the proof consists of having a uniform bound (valid for every domain Ω_γ with $\gamma \in V$) for weak solutions to the primal Boussinesq system, as discussed above. The boundary regularity of $(\boldsymbol{w}, S)$ is proved similarly as for the primal pair $(\boldsymbol{v}, T)$.

The next theorem is the main result of this chapter where we establish a differentiability result and a first-order optimality condition for problem (9).

Theorem 3 (Shape Derivative and Optimality Condition) *Let* $T_d \in H^2(\Omega^{\mathrm{M}}) \cap C_c(\Omega^{\mathrm{M}})$. *If* Re, Pr, Gr *satisfy* (21) *and* T_d *satisfies the smallness condition in Lemma 1, then* $\hat{J}$ *is directionally differentiable and its derivative is given by*

$$\hat{J}'(\gamma)h=-\int_\varepsilon^{1-\varepsilon}\big(F(\xi,\gamma(\xi))+\lambda\gamma''(\xi)\big)\,h(\xi)\,d\xi, \tag{22}$$

for every $\gamma, h \in V$, where $\varepsilon \in (0,1)$ is the one in the definition of V and

$$F(\xi, \gamma(\xi)) = \frac{1}{\mathrm{Re}}\left(\frac{\partial \boldsymbol{v}}{\partial \mathbf{n}} \cdot \frac{\partial \boldsymbol{w}}{\partial \mathbf{n}}\right)(\xi, \gamma(\xi)) + \frac{1}{\mathrm{RePr}}\left(\frac{\partial T}{\partial \mathbf{n}} \cdot \frac{\partial S}{\partial \mathbf{n}}\right)(\xi, \gamma(\xi))$$
$$+ \frac{1}{2}|T(\xi, \gamma(\xi)) - I(T, \gamma)|^2. \tag{23}$$

The pairs $(\boldsymbol{v}, T)$ and $(\boldsymbol{w}, S)$ are the weak solutions to the primal (2)–(7) and the adjoint (16)–(20) systems in Ω_γ, respectively. Moreover, if Ω_{γ^*} solves problem (9) with $\gamma^* \in V$ then

$$\int_{\varepsilon}^{1-\varepsilon} \left(F(\xi, \gamma^*(\xi)) + \lambda \gamma^{*\prime\prime}(\xi)\right)(\gamma^*(\xi) - h(\xi))\, d\xi \geq 0, \tag{24}$$

for all $h \in V$.

The proof (22) exploits the adjoint system given in (1) and follow from standard calculations following the adjoint method [17]. However, it is worth to mention that, at some point, we need to compute the derivative of the objective functional with respect to its second argument, γ, and this is where calculations benefit from working with domains with a unitary measure. The optimality condition (24) is then naturally deduced from (22).

4 Conclusions and Future Work

We presented a consistent functional approach to solve a shape optimization problem having a Boussinesq system as a constraint. We perturbed the bottom boundary of a two-dimensional unit square fulfilled with a viscous incompressible fluid, and look for an optimal domain to have a uniform temperature distribution. The fluid flow and temperature were described by a Boussinesq system with Dirichlet boundary conditions; modeling that there is no relative motion at the fluid-solid interface, and a heat source located at the bottom boundary part. The proposed geometrical setting allowed to prove the existence of an optimal domain as well as the derivation of an explicit formula for the directional derivatives of the reduced cost functional, leading to a first-order optimality condition. However, several results presented here are still true under milder regularity conditions on the bottom part of the boundary. In particular, the condition that every perturbed domain has a unitary volume was only required to simplify the directional derivative formula. However, this can be overcome through a regularization approach at the solution algorithm level. This is the next step in our research plan in the study of problem (9), along with investigating the performance of our theoretical results from numerical experiments. We further plan to extend our results to a more challenging problem, where mixed and (possibly) non-smooth boundary conditions are considered for the velocity field and temperature distribution of the fluid. We are particularly interested in considering do-nothing type boundary conditions, as in [2, 8, 9] (see also [3–5, 13, 15, 21, 22]). A detailed exposition of the results introduced in this chapter will be available soon.

Acknowledgments The authors acknowledge support from the American Institute of Mathematics (AIM) through the SQuaRE project *Optimal mixing and control of heat conductive flows*. Partial support for C.N.R. and L.M. was provided by NSF grants DMS-2012391 and DMS-2309557, respectively. W.H. received partial support from NSF grant DMS-2205117 and AFOSR grant FA9550-23-1-0675.

References

1. Adams, R.: Sobolev Spaces. Academic Press, New York (1975)
2. Arndt, R., Ceretani, A., Rautenberg, C.N.: On existence and uniqueness of solutions to a Boussinesq system with nonlinear and mixed boundary conditions. J. Math. Anal. Appl. **490**, 124201 (2020)
3. Beneš, M.: The "do nothing" problem for Boussinesq fluids. Appl. Math. Lett. **31**, 25–28 (2014)
4. Bertoglio, C., Caiazzo, A., Bazilevs, Y., Braack, M., Esmaily, M., Gravemeier, V., Marsden, A., Pironneau, O., Vignon-Clementel, I., Wall, W.: Benchmark problems for numerical treatment of backflow at open boundaries. Int. J. Numer. Methods Biomed. Eng. **34**(2), 2–34 (2018)
5. Bruneau, C.H., Fabrie, P.: New efficient boundary conditions for incompressible Navier-Stokes equations: a well posedness result. Math. Model. Numer. Anal. **30**(7), 815–840 (1996)
6. Burns, J., Hu, W.: Approximation methods for boundary control of the Boussinesq equations. In: Proceedings of the 52nd IEEE Conference on Decision and Control, pp. 454–459 (2013)
7. Burns, J., He, X., Hu, W.: Feedback stabilization of a thermal fluid system with mixed boundary control. Comput. Math. Appl. **71**(11), 2170–2191 (2016)
8. Ceretani, A., Rautenberg, C.N.: The Boussinesq system with mixed non-smooth boundary conditions and do-nothing boundary flow. Z. Angew. Math. Phys. **70**(14), 1–24 (2019)
9. Ceretani, A., Hu, W., Rautenberg, C.N.: Optimal conduit shape for Stokes flow. System Control Lett. **173**, 105461 (2023)
10. De Los Reyes, J.C., Tröltzsch, F.: Optimal control of the stationary Navier-Stokes equations with mixed control-state constraints. SIAM J. Control Optim. **46**(2), 287–309 (2007)
11. Evans, L.: Partial Differential Equations. American Mathematical Society, Providence (2010)
12. Galdi, G.P.: An Introduction to the Mathematical Theory of the Navier-Stokes Equations. Springer Science + Business Media, New York (2011)
13. Gresho, P.M.: Some current CFD issues relevant to the incompressible Navier-Stokes equations. Computer Methods Appl. Mech. Eng. **87**, 201–252 (1991)
14. Grisvard, P.: Elliptic Problems in Non-Smooth Domains. Pitman Publishing, Boston (1985)
15. Hintermüller, M., Kröner, A.: Differentiability properties for boundary control of fluid-structure interactions of linear elasticity with Navier-Stokes equations with mixed-boundary conditions in a channel. Appl. Math. Optim. **87**(2), 15 (2023)
16. Hintermüller, M., Rautenberg, C.N., Mohammadi, M., Kanitsar, M.: Boundary control for optimal mixing via Navier-Stokes flows. SIAM J. Optim. Control **55**(6), 3609–3639 (2017)
17. Hinze, M., Pinnau, R., Ulbrich, M., Ulbrich, S.: Optimization with PDE Constraints. Springer, New York (2008)
18. Hu, W., Wu, J.: Boundary control for optimal mixing via Navier-Stokes flows. SIAM J. Optim. Control **56**(4), 2768–2801 (2018)

19. Hu, W., Wu, J.: An approximating approach for boundary control of optimal mixing via Navier-Stokes flows. J. Differ. Equ. **267**, 5809–5850 (2019)
20. Kunisch, K., Marduel, X.: Optimal control of non-isothermal viscoelastic fluid flow. J. Non-Newton. Fluid Mech. **88**(3), 261–301 (2000)
21. Neustupa, T.: The weak solvability of the steady problem modelling the flow of a viscous incompressible heat-conductive fluid through the profile cascade. Int. J. Numer. Methods Heat Fluid Flow **27**(7), 1451–1466 (2017)
22. Sani, R.L., Gresho, P.M.: Résumé and remarks on the open boundary condition minisymposium. Int. J. Numer. Methods Fluids **18**(10), 983–1008 (1994)
23. Vieira, A., Cocquet, P.H.: The Boussinesq system with nonsmooth boundary conditions: existence, relaxation, and topology optimization. SIAM J. Optim. Control **60**(6), 3457–3484 (2022)

Toeplitz Subshifts, Equicontinuous Systems and Residually Finite Groups

María Isabel Cortez

Abstract This article is a short survey on Toeplitz subshifts and minimal equicontinuous systems, including characterizations of both systems and some of their dynamical properties. Additionally, we characterize infinite residually finite groups as those admitting Toeplitz subshifts and minimal equicontinuous Cantor systems with non-periodic points.

Keywords Toeplitz subshifts · Minimal equicontinuous systems · Residually finite groups

1 Introduction

Jacobs and Keane introduced Toeplitz subshifts in [24], in the context of $\mathbb{Z}$-actions. Since then, different works have shown that there are dynamical systems within this family of subshifts with varied behavior [4, 12, 14, 19, 23, 28, 31]. For example, in [13], Downarowicz shows that any possible set of invariant probability measures is realizable by some Toeplitz subshift in $\{0, 1\}^{\mathbb{Z}}$. The flexibility of Toeplitz subshifts motivated further generalizations beyond $\mathbb{Z}$-actions [6, 10, 26], providing examples of group actions on the Cantor set with interesting properties [11, 26]. Another consequence of the generalization of the Toeplitz subshift concept is the characterization of infinite residually finite groups as those admitting actions corresponding to non-periodic Toeplitz subshifts.

In this survey, we gather some results about Toeplitz subshifts and their relationship with minimal equicontinuous systems. Additionally, we recall some dynamical characterizations of residually finite groups involving both classes of systems. In the last section, we discuss some open questions related to Toeplitz subshifts and different aspects of topological dynamics and ergodic theory.

M. I. Cortez (✉)
Facultad de Matemáticas, Pontificia Universidad Católica de Chile, Edificio Rolando Chuaqui, Campus San Joaquín, Macul, Chile
e-mail: maria.cortez@uc.cl

P. E. Harris et al. (eds.), *Women in Mathematics in Latin America*,
Research Perspectives Ghent Analysis and PDE Center 14,
https://doi.org/10.1007/978-3-031-97857-9_7

2 Definitions and Background

In this article, by a *topological dynamical system* we mean a pair (X, G), where X is a compact metric space and G is a countable discrete group acting continuously on X. For every $x \in X$ and $g \in G$, we denote gx the action of g on x. The *stabilizer* of $x \in X$ is the subgroup $\mathrm{Stab}(x) = \{g \in G : gx = x\}$. We say that $x \in X$ is *non-periodic* if $\mathrm{Stab}(x) = \{1_G\}$, otherwise we say that x is *periodic*. A subset $A \subseteq X$ is *G-invariant* or just *invariant*, if $gA = A$ for every $g \in G$, where $gA = \{gx : x \in A\}$. The topological dynamical system (X, G) is *minimal* if for every $x \in X$ its orbit $O_G(x) = \{gx : g \in G\}$ is dense in X. We say that (X, G) is a *Cantor system* if X is a Cantor set. The topological dynamical system (X, G) is *equicontinuous* if the collection of homeomorphisms defined by the action of G on X, is an equicontinuous family. Let (X, G) and (Y, G) be topological dynamical systems. We say that $\pi : X \to Y$ is a *factor map* if π is surjective and $\pi(gx) = g\pi(x)$ for every $g \in G$ and $x \in X$. In this case, we say that (X, G) is a *topological extension* or an extension of (Y, G), and (Y, G) is a *topological factor* or a factor of (X, G). If π is bijective, we say that (X, G) and (Y, G) are *conjugate*. The factor map $\pi : X \to Y$ is *almost one-to-one* if the set $\{y \in Y : |\pi^{-1}\{\pi(y)\}| = 1\}$ is dense in Y. In this case, we say that (X, G) is an *almost 1-1 extension* of (Y, G). Every topological dynamical system (X, G) has a unique (up to conjugacy) *maximal equicontinuous factor*, that is, a factor (Y, G) which is equicontinuous, and such that every other factor from (X, G) to an equicontinuous system, passes through the factor from (X, G) to (Y, G). If (X, G) is a minimal almost 1-1 extension of an equicontinuous system (Y, G), then (Y, G) is the maximal equicontinuous factor of (X, G). See [1] for details about topological dynamical systems.

An *invariant probability measure* of the topological dynamical system (X, G) is a Borel probability measure μ of X such that $\mu(gA) = \mu(A)$, for every $g \in G$ and every Borel set $A \subseteq X$. This measure is *ergodic* if $\mu(A) \in \{0, 1\}$, for every invariant Borel set $A \subseteq X$. By $\mathcal{M}(X, G)$, we denote the space of Borel probability measures of (X, G). This space is convex, compact, and metrizable with respect to the weak* topology, and the extreme points of $\mathcal{M}(X, G)$ are precisely the ergodic measures of (X, G). It is well known that the group G is *amenable* if and only if for every topological dynamical system (X, G) the space $\mathcal{M}(X, G)$ is non-empty (see, for example, [25]).

3 Residually Finite Groups and Equicontinuous Actions on the Cantor Set

A group G is *residually finite* if for every $g \in G \setminus \{1_G\}$ there exist a finite group F and a homomorphism $\phi : G \to F$ such that $\phi(g) \neq 1_G$. Finite groups, abelian finitely generated groups, and finite rank-free groups are examples of residually finite groups. Divisible groups, as $\mathbb{Q}$, are not residually finite. The next characterization of residually finite groups is very well known. See [3, Section 2] for more details about these groups.

Proposition 1 *Let G be a countable group. Then the following conditions are equivalent:*

1. *G is residually finite;*
2. *There exists a decreasing sequence $(\Gamma_n)_{n\in\mathbb{N}}$ of finite index subgroups of G such that $\bigcap_{n\in\mathbb{N}}\Gamma_n = \{1_G\}$;*
3. *There exists a decreasing sequence $(\Gamma_n)_{n\in\mathbb{N}}$ of normal finite index subgroups of G such that $\bigcap_{n\in\mathbb{N}}\Gamma_n = \{1_G\}$.*

3.1 Odometers

Let G be a countable group. Let $(\Gamma_n)_{n\in\mathbb{N}}$ be a strictly decreasing sequence of finite index subgroups of G. The G-odometer associated to $(\Gamma_n)_{n\in\mathbb{N}}$ is defined as

$$X = \{(x_n)_{n\in\mathbb{N}} \in \prod_{n\in\mathbb{N}} G/\Gamma_n : \tau_n(x_{n+1}) = x_n, \text{ for every } n \in \mathbb{N}\},$$

where $\tau_n \colon G/\Gamma_{n+1} \to G/\Gamma_n$ is the canonical projection, for every $n \in \mathbb{N}$. The space X is a Cantor set if we endow every G/Γ_n with the discrete topology and $\prod_{n\in\mathbb{N}} G/\Gamma_n$ with the product topology. Let $\tau : G \to X$ given by $\tau(g) = (g\Gamma_n)_{n\in\mathbb{N}}$. The group G acts on X as $g(g_n\Gamma_n)_{n\in\mathbb{N}} = (gg_n\Gamma_n)_{n\in\mathbb{N}}$, for every $g \in G$ and $(g_n\Gamma_n)_{n\in\mathbb{N}} \in X$. If the groups $(\Gamma_n)_{n\in\mathbb{N}}$ are normal, then X is a subgroup of $\prod_{n\in\mathbb{N}} G/\Gamma_n$ and the map τ is a group homomorphism. The topological dynamical system (X, G) is known as the *G-odometer* or odometer associated to the sequence $(\Gamma_n)_{n\in\mathbb{N}}$. This system is equicontinuous and minimal. See [10, 14, 18], for more details on G-odometers.

The G-odometer (X, G) has a non-periodic point if and only if it is conjugate to a G-odometer associated to a decreasing sequence of finite index subgroups of G with trivial intersection (see, for example, [7]). This implies, according to Proposition 1, the following result:

Proposition 2 *Let G be a countable group. There exists a G-odometer with a non-periodic point if and only if G is residually finite.*

3.2 Automorphisms of Rooted Trees

Let $(\Sigma_n)_{n\in\mathbb{N}}$ be a sequence of finite sets with at least two elements. If we endow every Σ_n with the discrete topology, then $\Sigma = \prod_{n\geq 1}\Sigma_n$, with the product topology, is a Cantor set. This topology is induced by the metric given by

$$d((x_n)_{n\geq 1},(y_n)_{n\geq 1}) = \begin{cases} 0 & \text{if } x_n = y_n \text{ for all } n \geq 1 \\ \frac{1}{2^{k-1}} & \text{if } k = \min\{n \geq 1 : x_n \neq y_n\} \end{cases},$$

for every $(x_n)_{n\geq 1}$ and $(y_n)_{n\geq 1}$ in Σ. The *spherically homogeneous tree* associated to Σ is the infinite graph $T = (V, E)$, defined as follows: the set of vertices V is equal to the disjoint union $\bigcup_{n\geq 0} V_n$, where $V_0 = \{v_0\}$ and $V_n = \prod_{i=1}^{n} \Sigma_i$, for every $n \geq 1$. The set of edges E is equal to the disjoint union $\bigcup_{n\geq 1} E_n$, where E_n contains exactly one edge e from $v \in V_{n-1}$ to $va \in V_n$, for every $a \in \Sigma_n$ and $n \geq 1$. For every $e \in E$, we denote $s(e) \in V$ and $t(e) \in V$ the source and the target of e, respectively.

The set of infinite paths of T is given by

$$\partial T = \{(e_n)_{n\geq 1} \in \prod_{n\geq 1} E_n : t(e_n) = s(e_{n+1}), \forall n \geq 1\}.$$

The sequence $(x_n)_{n\geq 1} \in \Sigma$ is identified with the path $(e_n)_{n\geq 1} \in \partial T$ given by $t(e_1) = x_1$, $s(e_n) = x_1 \cdots x_{n-1}$ and $t(e_n) = x_1 \cdots x_n$, for every $n > 1$. This identification induces an isometry between Σ and ∂T, with the induced metric on ∂T.

An *automorphism* of T is a bijection $\tau : V \to V$ that preserves the structure of the tree, and such that the restriction $\tau|_{V_n}$ is a permutation of V_n, for every $n \geq 0$. More precisely, if $v \in V_n$ is equal to ua for $u \in V_{n-1}$ and $a \in \Sigma_n$, then there exists $b \in \Sigma_n$ such that $\tau(v) = \tau(u)b$, for every $n \geq 0$. The space of all the automorphisms of T, denoted by $\mathrm{Aut}(T)$, is a group with the composition of functions. Every element in $\mathrm{Aut}(T)$ induces an isometry on ∂T, and conversely, every isometry in ∂T induces an automorphism in $\mathrm{Aut}(T)$.

The next result says that every countable residually finite group G is isomorphic to a subgroup Γ of $\mathrm{Aut}(T)$, for some tree T, such that the action of Γ on ∂T is isomorphic to a G-odometer. Different versions of these result can be found in [22, Remark 1], [21, Proposition 2.8] and [8].

Proposition 3 *Let G be a countable residually finite group with a decreasing sequence $(\Gamma_n)_{n\geq 1}$ of finite index subgroups of G with trivial intersection. Let (X, G) be the G-odometer associated to $(\Gamma_n)_{n\geq 1}$ and let T be the spherically homogeneous tree associated to $(\Sigma_n)_{n\geq 1}$, where $\Sigma_n = \{0, \cdots, [G : \Gamma_n] - 1\}$, for every $n \geq 1$. Then there exists an injective homomorphism $\psi : G \to \mathrm{Aut}(T)$ and an isometry $\phi : X \to \partial T$, such that $\phi(gx) = \psi(g)\phi(x)$ for every $g \in G$ and $x \in X$.*

Proof If $T = (V, E)$, with $V = (V_n)_{n\geq 0}$, we can identify G/Γ_n with V_n, for every $n \geq 1$. There is an edge from $a\Gamma_n \in V_n$ to $b\Gamma_{n+1} \in V_{n+1}$ if $b\Gamma_n = a\Gamma_n$. This identification induces a natural bijective isometry $\phi : X \to \partial T$. The map $\psi : G \to \mathrm{Aut}(T)$ given by $\psi(g) = \phi \circ g \circ \phi^{-1}$ is a well defined homomorphism, and the trivial intersection of the Γ_n's implies that ψ is injective. The $\psi(G)$ action on ∂T is isomorphic to the G-odometer through the function ϕ. □

3.3 Characterization of Equicontinuous Actions on the Cantor Set

Proposition 4 *The system (X, G) is a minimal equicontinuous Cantor system if and only if (X, G) is conjugate to a G-odometer.*

Proof It is sufficient to follow the proof of [9, Theorem 2.7], and to note that if (X, G) is not free, then the function $\varphi : G \to K$ is not necessarily injective. This implies that the action of G on K is not necessarily free. The intersection of the groups G_n's mentioned there, is the stabilizer of the neutral element $e \in K$ (and maybe non-trivial). Then, the proof follows to conclude that (X, G) is regularly recurrent, meaning that for every $x \in X$ and every clopen neighborhood C of x, there exists a finite index subgroup Γ of G such that $\gamma x \in C$, for every $\gamma \in \Gamma$. Let $x \in X$ and $(U_n)_{n\in\mathbb{N}}$ be a nested sequence of clopen sets such that $\bigcap_{n\in\mathbb{N}} U_n = \{x\}$. Then, by [10, Lemma 4], for every $n \in \mathbb{N}$ there exists a finite index subgroup Γ_n such that $C_n = \overline{O_{\Gamma_n}(x)} \subseteq U_n$ is a clopen set verifying $C_n \cap \gamma C_n \neq \emptyset$ if and only if $\gamma C_n = C_n$, and if and only if $\gamma \in \Gamma_n$. The system (X, G) is conjugate to the odometer given by the sequence $(\Gamma_n)_{n\in\mathbb{N}}$ (see the proof of [10, Theorem 2]). □

The following result is a direct consequence of Propositions 2, 3, and 4.

Proposition 5 *Let G be an infinite countable group. The following statements are equivalent:*

1. *G is residually finite;*
2. *There exists a G-odometer with a non-periodic point;*
3. *There exists a spherically homogeneous tree T such that G is isomorphic to a subgroup of* $\mathrm{Aut}(T)$ *whose action on ∂T has a non-periodic point;*
4. *There exists a minimal equicontinuous Cantor system (X, G) with a non-periodic point.*

4 Residually Finite Groups and Subshifts

Let A be a finite set with at least two elements. Consider the set

$$A^G = \{x = (x(g))_{g\in G} : x(g) \in \Sigma, \text{ for every } g \in G\}.$$

The space A^G is a Cantor set when we endow A with the discrete topology and A^G with the product topology. The group G acts continuously on A^G by means of the *shift action*, which is defined as follows: let $x = (x(g))_{g\in G} \in A^G$ and $h \in G$,

$$hx(g) = x(h^{-1}g), \text{ for every } g \in G.$$

The Cantor system (A^G, G) is called the $|A|$-full G-shift, where $|A|$ denotes the cardinality of A. A *G-subshift* or *subshift* of A^G is a closed G-invariant subset $X \subseteq$

A^G. We also call G-subshift or subshift the topological dynamical system (X, G) given by the restriction of the shift-action of G on X. The class of infinite subshifts is disjoint from the class of equicontinuous systems. Nevertheless, residually finite groups are also characterized by the dynamical properties of its subshifts. The following result can be found in [3].

Theorem 1 *The group G is residually finite if and only if the set of points of $\{0, 1\}^G$ which have a finite orbit with respect to the shift action, is dense in $\{0, 1\}^G$.*

Proof The idea of the proof is based on the fact that x has a finite G-orbit if and only if $\mathrm{Stab}(x)$ is a finite index subgroup of G. See [3, Theorem 2.7.1] □

4.1 Toeplitz Subshifts

Infinite subshifts with a dense set of points having finite orbits are not minimal. Unlike Theorem 1, in this section, we characterize residually finite groups through a particular family of minimal subshifts. With this purpose, for every $x \in A^G$ and every finite index subgroup Γ of G, we define

$$\mathrm{Per}(x, \Gamma, \alpha) = \{g \in G : x(\gamma g) = \alpha \text{ for every } \gamma \in \Gamma\}, \text{ for every } \alpha \in A,$$

$$\mathrm{Per}(x, \Gamma) = \bigcup_{\alpha \in A} \mathrm{Per}(x, \Gamma, \alpha).$$

An element $x \in A^G$ is *Toeplitz* if for every $g \in G$ there exists a finite index subgroup Γ of G such that $g \in \mathrm{Per}(x, \Gamma)$. A subshift $X \subseteq A^G$ is a *Toeplitz subshift* if there exists a Toeplitz element $x \in A^G$, such that $X = \overline{O_G(x)}$. Since every Toeplitz element is almost periodic, Toeplitz subshifts are always minimal (see [10, 26, 27]). Observe that if $x \in A^G$ is a non-periodic Toeplitz element, then the intersection of all the finite index subgroups Γ of G such that $\mathrm{Per}(x, \Gamma) \neq \emptyset$, is trivial. Thus, the existence of non-periodic Toeplitz sequences in A^G implies that G is residually finite. The next result shows that this is an equivalence.

Proposition 6 *Let G be an infinite countable group. Then G is residually finite if and only if there exists a Toeplitz subshift $X \subseteq \{0, 1\}^G$ with no periodic points.*

Proposition 6 is a consequence of the following theorem, which corresponds to [10, Theorem 7].

Theorem 2 *Let G be a countable group. For every minimal equicontinuous Cantor system (Z, G) there exists a Toeplitz subshift $X \subseteq \{0, 1\}^G$ such that (X, G) is an almost 1-1 extension of (Z, G).*

4.2 Toeplitz Subshifts and Their Maximal Equicontinuous Factors

In this section, we will see the converse of Theorem 2, i.e., that every Toeplitz subshift is an almost 1-1 extension of some minimal equicontinuous Cantor system. Consequently, we will characterize Toeplitz subshifts as the minimal symbolic almost 1-1 extensions of odometers.

Since G is countable, we have that $x \in A^G$ is a Toeplitz element if and only if there exists a decreasing sequence $(\Gamma_n)_{n\in\mathbb{N}}$ of finite index subgroups of G such that $G = \bigcup_{n\in\mathbb{N}} \mathrm{Per}(x, \Gamma_n)$. Moreover, it is possible to choose the sequence $(\Gamma_n)_{n\in\mathbb{N}}$ as a *period structure* (see [10] or [5]), i.e., such that for every $n \in \mathbb{N}$, the group Γ_n satisfies the following additional property: if $\mathrm{Per}(x, \Gamma_n, \alpha) \subseteq \mathrm{Per}(gx, \Gamma_n, \alpha)$ for every $\alpha \in A$, then $g \in \Gamma_n$.

Proposition 7 *Let $x \in A^G$ be a Toeplitz element and $X = \overline{O_G(x)} \subseteq A^G$ be the associated Toeplitz subshift. If $(\Gamma_n)_{n\in\mathbb{N}}$ is a period structure of x, then the maximal equicontinuous factor of (X, G) is the G-odometer associated to $(\Gamma_n)_{n\in\mathbb{N}}$.*

Proof If $(\Gamma_n)_{n\in\mathbb{N}}$ is a period structure of x, then for every $n \in \mathbb{N}$, the set of return times of C_n, with

$$C_n = \{y \in X = \overline{O_G(x)} : \mathrm{Per}(y, \Gamma_n, \alpha) = \mathrm{Per}(x, \Gamma_n, \alpha), \text{ for every } \alpha \in \Sigma\},$$

is exactly Γ_n. More precisely, for every $g, h \in G$ we have $gC_n = hC_n$ if and only if $g\Gamma_n = h\Gamma_n$, which is also equivalent to $gC_n \cap hC_n \neq \emptyset$ (see [10, Lemma 8 and Proposition 6] or [5]). This implies that $(\mathcal{P}_n = \{gC_n : g \in D_n\} : n \in \mathbb{N})$ is a nested sequence of clopen partitions of X, where $D_n \subseteq G$ is any set containing exactly one element of each class in $\{g\Gamma_n : g \in G\}$, for every $n \in \mathbb{N}$. Since $\{x\} = \bigcap_{n\in\mathbb{N}} C_n$, the maps $\phi_n : X \to G/\Gamma_n$ given by $\phi_n(y) = v\Gamma_n$ if and only if $y \in vC_n$, define an almost 1-1 factor map from (X, G) to the G-odometer associated to $(\Gamma_n)_{n\in\mathbb{N}}$. This implies the result. □

The following result summarizes the relationship between minimal equicontinuous Cantor systems and Toeplitz subshifts.

Proposition 8 *Let (X, G) be a minimal subshift. The following statements are equivalent:*

1. *(X, G) is a Toeplitz subshift;*
2. *There exists a minimal equicontinuous Cantor system (Z, G) such that (X, G) is an almost 1-1 extension of (Z, G).*

Moreover, if $\pi : X \to Z$ is the almost 1-1 factor map, then

$$\{x \in X : x \text{ is Toeplitz}\} = \pi^{-1}\{z \in Z : |\pi^{-1}\{z\}| = 1\}.$$

Proof From Proposition 7, we get 1. implies 2. From [10, Proposition 5, Theorem 2] and Proposition 4, we get 2. implies 1., and the fact that the set of Toeplitz elements coincides with the pre-images of points with only one pre-image. □

4.3 Toeplitz Subshifts and Realization Results

Among several interesting results, Williams shows in [31] there exist Toeplitz subshifts in $\{0, 1\}^{\mathbb{Z}}$ with an arbitrary topological entropy in $[0, \log 2)$. Later, in [13] Downarowicz proves that any metrizable Choquet simplex is realizable as the set of invariant probability measures of a Toeplitz subshift. Both results were extended for Toeplitz subshifts given by the actions of amenable residually finite groups, mentioned below.

Theorem 3 ([26, Theorem 1.1]) *Let G be an infinite, countable, amenable, and residually finite group. For every finite set A with at least two elements, and every $t \in [0, \log|A|)$, there exists a Toeplitz subshift $X \subseteq A^G$ with topological entropy equal to t.*

Theorem 4 ([11, Theorem A]) *Let G be an infinite, countable, amenable, and residually finite group. For every metrizable Choquet simplex K and any G-odometer (Z, G) associated to a decreasing sequence of normal finite index subgroup of G with trivial intersection, there exists a Toeplitz G-subshift which is an almost 1-1 extension of (Z, G) such that the set of its invariant probability measures is affinely homeomorphic to K.*

Theorem 4 prescribes at the same time the odometer and the simplex of invariant measures, which implies, at least for $\mathbb{Z}$-actions, that there are uncountable many non-orbit equivalent Toeplitz subshifts with the same simplex of invariant measures (see [16] for the relation between orbit equivalence and dimension group, and [28, 29] for similar results related to the prescription of dimension groups). Still, in the context of $\mathbb{Z}$-actions, Hoynes shows in [23] that it is also possible to prescribe, at the same time, the topological entropy. Applying the characterization of orbit equivalence for minimal free $\mathbb{Z}^d$-actions on the Cantor set given by Giordano et al. in [17], we get the following result:

Theorem 5 ([11, Theorem B]) *For every integer $d > 1$ and every Toeplitz $\mathbb{Z}$-subshift $(X, \mathbb{Z})$, there exists a Toeplitz $\mathbb{Z}^d$-subshift which is orbit equivalent to $(X, \mathbb{Z})$.*

In the context of non-amenable groups and the realization of simplices of invariant measures, there are the following results (see [31] for a version of the first theorem, in the context of $\mathbb{Z}$-actions):

Theorem 6 ([5, Theorem 1.3]) *Let G be a countable residually finite group and let (Z, G) be a G-odometer associated to a decreasing sequence of normal finite index subgroup of G with trivial intersection. For every integer $r > 1$ there exists* *a Toeplitz G-subshift with at least r ergodic probability measures* $\nu_1, \dots, \nu_r$*, and whose maximal equicontinuous factor is* (Z, G)*. Furthermore, for every* $1 \leq i \leq r$*, the system* (X, σ, G, ν_i) *is measure-theoretic conjugate to* (Z, G, ν)*, where* ν *is the Haar measure of* Z*.*

Theorem 7 ([7]) *Let G be a non-amenable residually finite group. Then for every minimal equicontinuous Cantor system* (Z, G) *having a point with trivial stabilizer, there exists a Toeplitz G-subshift with no invariant probability measures, whose maximal equicontinuous factor is* (Z, G)*.*

5 Final Discussion and Questions

It is natural to wonder whether the realization results can be extended to the Toeplitz subshifts given by the actions of non-amenable groups. Some obstructions are already known. For example, Glasner and Weiss show in [20] that if G has the property T, then the space of invariant measures of a topological dynamical system (X, G) is a simplex whose set of extreme points is closed. Thus, if G has the property T, the Poulsen simplex cannot be realized as the space of invariant probability measures of topological dynamical systems (X, G). On the other hand, the invariant measures of the Toeplitz subshifts constructed in [5] are approximated by periodic measures, so we wonder if this is true for every Toeplitz G-subshift, even if the group G does not have the PA property (see [2] for details about PA property on groups). It is also well-known that if G is non-amenable if and only if a topological dynamical system (X, G) with no invariant measures exists. Thus, Theorem 7 provides a characterization of amenability for residually finite groups. It is important to mention that this result was obtained also by Tsankov [30]. In [15], the authors show that the $0 - 1$ subshifts are a test family for the amenability of countable groups, which makes it natural to wonder whether the same is true for Toeplitz subshifts and non-amenable residually finite groups.

There are orbit equivalent $\mathbb{Z}$ and $\mathbb{Z}^2$-odometers conjugated to actions on the boundary of the same spherically homogeneous tree. Thus, it is natural to wonder if, given a group acting on the boundary of a rooted tree, there exists $\mathbb{Z}$-action with the same orbits.

? Questions

1. Let G be a non-amenable residually finite group. What are the Choquet simplices that are affine homeomorphic to the space of invariant probability measures of a G-Toeplitz subshift?
2. Are the 0-1 Toeplitz subshifts a test family for the amenability of residually finite groups?

(continued)

3. Let G be a non-amenable residually finite group. Is it possible to approximate all the invariant measures of a Toeplitz G-subshift by periodic measures?
4. Is it possible to construct explicit examples of orbit equivalence maps and algebraic invariants?

References

1. Auslander, J.: Minimal Flows and Their Extensions. North-Holland Mathematics Studies, vol. 153. North-Holland, Amsterdam (1988)
2. Bowen, L.: Periodicity and circle packings of the hyperbolic plane. Geom. Dedicata **102**, 213–236 (2003)
3. Ceccherini-Silberstein, T., Coornaert, M.; Cellular Automata and Groups. Springer, Berlin, Heidelberg (2010)
4. Cecchi Bernales, P., Donoso, S.: Strong orbit equivalence and superlinear complexity. Israel Journal of Mathematics, to appear
5. Cecchi Bernales, P., Cortez, M.I., Gómez, J.: Invariant measures of Toeplitz subshifts on non-amenable groups. Ergodic Theory Dyn. Syst. **44**(11), 3186–3215 (2024)
6. Cortez, M.I.: $\mathbb{Z}^d$-Toeplitz arrays. Discrete Contin. Dyn. Syst. **15**(3), 859–881 (2006)
7. Cortez, M.I., Gómez, J.: Almost 1-1 extensions of Furstenberg-Weiss type and measures. Transaction of the American Mathematical Society, to appear
8. Cortez, M.I., Lukina, O.: Settled elements in profinite groups. Adv. Math. **404**(Part B), 108424 (2022)
9. Cortez, M.I., Medynets, K.: Orbit equivalence rigidity of equicontinuous systems. J. Lond. Math. Soc. (2) **94**(2), 545–556 (2016)
10. Cortez, M.I., Petite, S.: G-odometers and their almost one-to-one extensions. J. Lond. Math. Soc. (2) **78**(1), 1–20 (2008)
11. Cortez, M.I., Petite, S.: Invariant measures and orbit equivalence for generalized Toeplitz subshifts. Groups Geom. Dyn. **8**, 1007 (2014)
12. Donoso, S., Durand, F., Maass, A., Petite, S.: On automorphism groups of Toeplitz subshifts. Discrete Anal., Paper No. 11, 19 pp. (2017)
13. Downarowicz, T.: The Choquet simplex of invariant measures for minimal flows. Israel J. Math. **74**(2–3), 241–256 (1991)
14. Downarowicz, T.: Survey of odometers and Toeplitz flows. In: Algebraic and Topological Dynamics. Contemp. Math., vol. 385, pp. 7–37. American Mathematical Society, Providence (2005)
15. Frisch, J., Kechris, A., Shinko, F., Vidnyánszky, Z.: Realizations of countable Borel equivalence relations. Preprint. arXiv:2109.12486 (2021)
16. Giordano, T., Putnam, I.F., Skau, C.F.: Topological orbit equivalence and C^*-crossed products. J. Reine Angew. Math. **469**, 51–111 (1995)
17. Giordano, T., Matui, H., Putnam, I.F., Skau, C.F.: Orbit equivalence for Cantor minimal $\mathbb{Z}^d$-systems. Invent. Math. **179**(1), 119–158 (2010)
18. Giordano, T., Putnam, I.F., Skau, C.F.: $\mathbb{Z}^d$-odometers and cohomology. Groups Geom. Dyn. **13**(3), 909–938 (2019)
19. Gjerde, R., Johansen, O.: Bratteli-Vershik models for Cantor minimal systems: applications to Toeplitz flows. Ergodic Theory Dyn. Syst. **20**(6), 1687–1710 (2000)

20. Glasner, E., Weiss, B.: Kazhdan's property T and the geometry of the collection of invariant measures. Geom. Funct. Anal. **7**(5), 917–935 (1997)
21. Grigorchuk, R.I.: Some problems of the dynamics of group actions on rooted trees (Russian). Tr. Mat. Inst. Steklova **273**, 72–191 (2011), Sovremennye Problemy Matematiki; translation in Proc. Steklov Inst. Math. **273**(1), 64–175 (2011)
22. Grigorchuk, R., Šuniç, Z.: Self-similarity and Branching in Group Theory. Groups St. Andrews 2005. Vol. 1, pp. 36–95. London Math. Soc. Lecture Note Ser., vol. 339. Cambridge University Press, Cambridge (2007)
23. Hoynes, S.M.: Toeplitz flows and their ordered K-theory. Ergodic Theory Dyn. Syst. **36**(6), 1892–1921 (2016)
24. Jacobs, K., Keane, M.: $0-1$-sequences of Toeplitz type. Z. Wahrsch. Verw. Geb. **13**, 123–131 (1969)
25. Kerr, D., Li, H.: Ergodic Theory. Springer Monogr. Math. Springer, Cham (2016)
26. Krieger, F.: Sous-décalages de Toeplitz sur les groupes moyennables résiduellement finis. J. Lond. Math. Soc. (2) **75**(2), 447–462 (2007)
27. Krieger, F.: Toeplitz Subshifts and Odometers for Residually Finite Groups. Sémin. Congr., vol. 20, pp. 147–161. Société Mathématique de France, Paris (2010)
28. Sugisaki, F.: Toeplitz flows, ordered Bratteli diagrams and strong orbit equivalence. Ergodic Theory Dyn. Syst. **21**(6), 1867–1881 (2001)
29. Sugisaki, F.: On the subshift within a strong orbit equivalence class for minimal homeomorphisms. Ergodic Theory Dyn. Syst. **27**(3), 971–990 (2007)
30. Tsankov, T.: Personal communication
31. Williams, S.: Toeplitz minimal flows which are not uniquely ergodic. Z. Wahrsch. Verw. Geb. (1) **67**, 95–107 (1984)

Slowly Non-dissipative Reaction–Diffusion Equations with a Jumping Nonlinearity

Julia do Prado Pazzini, Juliana Fernandes da Silva, and Juliana Honda Lopes

Abstract The aim of this paper is to address the asymptotic behavior of a reaction-diffusion equation driven by a jumping nonlinearity. Providing some conditions in the nonlinear terms, we ensure that the equation is a slowly non-dissipative equation, i.e., blow-up occurs but only in infinite time.

Keywords Reaction-diffusion equations · Jumping nonlinearity · Slowly non-dissipative systems · Blow-up solutions

1 Introduction

In the last few decades, the study of longtime behavior of dynamical systems have been widely investigated not only for those generated by equations whose solutions remain uniformly (with time) bounded in a set, in which there is a set of bounded solutions, called the global attractor, but also for those that have a subset of solutions blowing-up at infinite time.

When dealing with solutions of reaction-diffusion equations, there are three terminologies depending on their asymptotic behaviors: dissipative, fast non-dissipative or slowly non-dissipative. In the former case, all the trajectories for the dynamical system will be eventually bounded; in the last two cases, blow-up in time should occur, either in finite time or just at an infinite time, respectively.

This study will concentrate on the case of the latter type of systems. More specifically, we seek to identify a more general class of equations, broader than those discussed in [1], that similarly produces a slowly non-dissipative dynamical

---

J. d. P. Pazzini · J. Fernandes da Silva (✉)
Instituto de Matemática da Universidade Federal do Rio de Janeiro, CT - Centro de Tecnologia, Cidade Universitária, Rio de Janeiro, Brazil
e-mail: jfernandes@im.ufrj.br

J. H. Lopes
Instituto de Matemática e Estatística, Universidade de São Paulo, São Paulo, Brazil

P. E. Harris et al. (eds.), *Women in Mathematics in Latin America*,
Research Perspectives Ghent Analysis and PDE Center 14,
https://doi.org/10.1007/978-3-031-97857-9_8

systems. We will demonstrate that the following class of reaction-diffusion equations fulfills these criteria. Consider

$$\begin{cases} u_t = u_{xx} + b^+u^+ - b^-u^- + g(u), & x \in [0, \pi], \\ u(t, 0) = u(t, \pi) = 0, \end{cases} \tag{1}$$

where g is a bounded and locally Lipschitz function, b^+ and b^- are real values, $u^+ = max\{u, 0\}$ and $u^- = max\{-u, 0\}$, i.e.,

$$f(x, u) = b^+u^+ - b^-u^- + g(u)$$

is a nonlinear locally Lipschitz and continuous function with a positive growth and a bounded nonlinear term. The aim of this paper is to prove that (1) will produce grow-up solutions, which are solutions that blow-up in infinite time.

Initial value problems involving jumping nonlinearities were first studied by Dancer and Fučik (see, for example, [2, 3]), and have been analyzed through several perspectives since then. Suspension bridge is an example of dynamical system with jumping nonlinearity, which carries the unexpected difficulties for the system such as, for an extensive range of initial conditions, a given push can produce either small or large oscillations on the bridge, see [4]. From the dynamics point of view, the case $b^+ = b^-$ was treated in [1] and [5], where a description of the related global attractor is provided. See [6] for a generalization to quasilinear equations.

We focus on presenting the more general class of equations as in (1), which appears to be significantly richer from the perspective of asymptotic dynamics. Our work sheds some light in that direction and it is divided as follows. In Sect. 2, we establish some notations and introduce some preliminary concepts that will be used throughout the text. In Sect. 3, we prove the existence and uniqueness of solutions for Eq. (1). Finally, through eigendecomposition, we prove our main result regarding the growth of solutions for (1), and confirming that we are indeed in the realm of slowly non-dissipative systems.

2 Preliminaries

Throughout this paper, we work with the Hilbert spaces $L^2([0, \pi])$ with norm $\|\cdot\|_{L^2([0,\pi])}$ and inner product $\langle\cdot,\cdot\rangle_{L^2([0,\pi])}$, and $H^1([0, \pi])$ with norm $\|\cdot\|_{H^1([0,\pi])}$.

Before verifying that the dynamical system generated by (1) is slowly non-dissipative, it is important to clarify to be more precise about the terminology and differences between the three possible behaviors, which are dissipative, fast non-dissipative and slowly non-dissipative.

A dynamical system is known as dissipative if in its state space there exists a bounded region where every solution enters and continues there for all time. If the solutions do not exist for all time and blow-up in finite time, then the system is called fast non-dissipative. Finally, if the dynamical system has trajectories that exist for all time and, as well, has at least one solution that blows up at infinite time, then it is slowly non-dissipative and that specific solution is known as grow-up solution.

When studying the asymptotic behavior of equations, a fundamental object to be described is the global attractor, which, in the dissipative setting, has a characterization that is standard in the literature. Briefly, the global attractor is a maximal compact invariant set that attracts all bounded sets.

In non-dissipative settings, the existence of solutions exhibiting blow–up behavior in time, requires the dropping of the compactness assumption. This led to the introduction of the concept of non-compact attractors, as defined in [7], see also [8]. More precisely, we have

Definition 1 A non-compact global attractor is a closed non-empty minimal set in X attracting all bounded sets of X.

3 Slowly Non-dissipativeness

Our first goal is to ensure global existence in time of solutions for Eq. (1).

Lemma 1 *Equation (1) generates a semiflow on $H^1([0,\pi])$, i.e., for all $u_0 \in H^1([0,\pi])$, there exists a unique $u(t,\cdot) \in H^1([0,\pi])$, $t \in [0,+\infty)$, with $u(0,\cdot) = u_0$.*

Proof To ensure the local existence of solution for the problem (1) and that this equation generates a semiflow in $H^1([0,\pi])$, let us apply Theorem 3.3.3 of [9].

The Laplace operator $-\partial_{xx}$ being sectorial, we are left with only proving that

$$\begin{aligned} f : H^1([0,\pi]) &\longrightarrow L^2([0,\pi]) \\ u &\longmapsto b^+u^+ - b^-u^- + g(u), \end{aligned}$$

is locally Lipschitz. In fact, from the hypothesis on g, we get

$$\begin{aligned} \|f(u_1)-f(u_2)\|_{L^2([0,\pi])} &\le c(|b^+|+|b^-|)\|u_1-u_2\|_{H^1([0,\pi])}+L\|u_1-u_2\|_{H^1([0,\pi])} \\ &\le M\|u_1-u_2\|_{H^1([0,\pi])}, \end{aligned}$$

where the Lipschitz constant M depends on b^+, b^- and the Lipschitz constant of g. Then, local existence of solutions is guaranteed.

Now, to ensure the global existence, the idea is to apply Corollary 3.3.5 of [9]. Using the fact that g is bounded, we can define a constant k as the maximum between the upper bound of g, $|b^+|$ and $|b^-|$. Then, for all $u \in H^1([0,\pi])$, we get

$$\begin{aligned}\|f(u)\|_{L^2([0,\pi])} &\leq \max\{|b^+|, |b^-|\}\|u^+ - u^-\|_{H^1([0,\pi])} + c \\ &\leq k\left(1 + \|u\|_{H^1([0,\pi])}\right).\end{aligned}$$

Therefore, Eq. (1) defines a semiflow on $H^1([0,\pi])$. □

To try to understand the limit-behavior of the solutions of the problem (1), it is necessary to study the following problem:

$$\begin{cases}-u_{xx} = b^+u^+ - b^-u^- + \lambda u \\ u(t,0) = u(t,\pi)\end{cases}. \tag{2}$$

There are several works related to the problem (2), see for example [10]. The points λ for which (2) has a non-trivial solution are called *half-eigenvalues* and the solutions are called *half-eigenfunctions*. However, the theory of these half-eigenfunctions is not yet completely understood. It is then more convenient to use the eigenfunctions of the operator $A_b = -\partial_{xx} + bI$ for some $b > 0$. For this purpose, let us consider the following equivalent problem: since $u = u^+ - u^-$, (1) can be rewritten as

$$u_t = u_{xx} + b_{\max}u - (b_{\max} - b_{\min})\,u^* + g(u), \tag{3}$$

where $b_{\max} := \max\{b^+, b^-\}$, $b_{\min} := \min\{b^+, b^-\}$ and

$$u^* := \begin{cases}u^+, \text{ if } b_{\max} = b^- \\ -u^-, \text{ if } b_{\max} = b^+\end{cases}.$$

The right side of (3) consists of a linear part, $u_{xx} + b_{\max}u$ and of a non-linear part, $F(u) := -(b_{\max} - b_{\min})u^* + g(u)$. The operator $A_{b_{\max}} := -\partial_{xx} - b_{\max}I$ is a sectorial operator which generates an analytic semigroup $e^{-tA_{b_{\max}}}$. The operator $A_{b_{\max}}$ with Dirichlet boundary conditions generates an orthonormal basis $\{\phi_j(\cdot)\}_{j\in\mathbb{N}}$ of $L^2([0,\pi])$, given by $\phi_0(x) = \sqrt{\frac{1}{\pi}}$, $\phi_j(x) = \sqrt{\frac{2}{\pi}}\sin(jx)$, $j = 1, 2, \cdots$. Let us denote by λ_j the corresponding eigenvalues, which are given by $\lambda_j = (j+1)^2 - b_{\max}$, for each $j \in \mathbb{N}$.

The following is our main result, which provides necessary conditions in the nonlinear term so that Eq. (3) generates a slowly non-dissipative dynamical system.

Theorem 1 *For $b_{\max} > 1$, i.e., the first eigenvalue of the operator $-\partial_{xx}$, Eq. (3) generates a slowly non-dissipative dynamical system.*

Proof Let us consider the eigenspace E_j associated with each eigenvalue λ_j of $A_{b_{\max}}$. Given any solution $u(t,x)$ of (1) and denoting by $\widehat{u}_j(t) = \langle u(t,\cdot), \phi_j(\cdot)\rangle_{L^2([0,\pi])}$, then using the L^2-eigenprojection, it can be represented by $u(t,x) = \sum_{j=0}^{\infty} \widehat{u}_j(t)\phi_j(x)$. Applying the E_j-projection in Eq. (3), we get

$$\begin{aligned}\frac{d}{dt}\widehat{u}_j(t) =& \frac{d}{dt}\langle u(t,\cdot), \phi_j(\cdot)\rangle_{L^2([0,\pi])}\\ =& \langle u_{xx} + b_{\max}u, \phi_j(\cdot)\rangle_{L^2([0,\pi])} + \langle (b_{\min} - b_{\max})u^* + g(u), \phi_j(\cdot)\rangle_{L^2([0,\pi])},\end{aligned}$$

which implies

$$\frac{d}{dt}\widehat{u}_j(t) = -\lambda_j\widehat{u}_j(t) + \langle (b_{\min} - b_{\max})u^* + g(u), \phi_j(\cdot)\rangle_{L^2([0,\pi])}. \tag{4}$$

Denoting by $\langle F(u)(\cdot), \phi_j(\cdot)\rangle_{L^2([0,\pi])} := \langle (b_{\min} - b_{\max})u^* + g(u), \phi_j(\cdot)\rangle_{L^2([0,\pi])}$, where $F(u)$ is the Nemytskii operator for $(b_{\min} - b_{\max})u^* + g(u)$, and denoting by $\widehat{f}_j(t) := \langle F(u(t,\cdot))(\cdot), \phi_j(\cdot)\rangle_{L^2([0,\pi])}$, we finally rewrite (4) as

$$\frac{d}{dt}\widehat{u}_j(t) = -\lambda_j\widehat{u}_j(t) + \widehat{f}_j(t). \tag{5}$$

We observe that if $b_{\max} > 1$, then $\lambda_j = (j+1)^2 - b_{max} < 0$, at least for $j = 0$.

Since it is enough to exhibit at least one solution that blow-up when $t \to +\infty$, let us suppose without loss of generality that $b_{\max} = b^+$ and that there exist a positive solution of (3). In fact, we can choose $u_0 = \sin x$, then by the fact that the zero-number is a non-increasing function (also known as Matano's Principle), the solution of (3) related to this initial condition will be a positive function. Therefore, in this case, $F(u) = g(u)$, which is uniformly bounded in u. Then (5) can be rewritten as

$$\frac{d}{dt}\widehat{u}_j(t) = -\lambda_j\widehat{u}_j(t) + \widehat{g}_j(t), \tag{6}$$

where $\widehat{g}_j(t) := \langle G(u(,\cdot))(\cdot), \phi_j(\cdot)\rangle_{L^2([0,\pi])}$ and $G(u)$ is the Nemytskii operator of $g(u)$. Then the non-homogeneous ODE (6) has a solution of the form $\widehat{u}_j(t) = \widehat{u}_j^p(t) + \widehat{u}_j^h(t)$, where $\widehat{u}_j^p(t)$ is a particular solution of (6) and $\widehat{u}_j^h(t)$ is a solution of the corresponding homogeneous equation. As a particular solution, we write

$$\widehat{u}_j^p(t) = \int_{\infty}^{t} e^{-\lambda_j(t-s)}\widehat{g}_j(s)\,ds,$$

and the solution with initial condition

$$\widehat{u}_j(0) = \int_{\infty}^{0} e^{\lambda_j s}\widehat{g}_j(s)\,ds + \widehat{u}_j^h(0),$$

is given by

$$\widehat{u}_j(t) = \widehat{u}^h_j(0)e^{-\lambda_j t} + \int_\infty^t e^{-\lambda_j(t-s)}\widehat{g}_j(s)\,ds.$$

Knowing that $G(u)$ is bounded in $L^2([0,\pi])$, then $\widehat{g}_j(t)$ is uniformly bounded for all $j \in \mathbb{N}_0$. Consequently, the particular solution $\widehat{u}^p_j(t)$ is uniformly bounded for all $j \in \mathbb{N}_0$ such that $\lambda_j < 0$. On the other hand, when $\lambda_j < 0$ and $\widehat{u}^h_j(0) \neq 0$, then $\widehat{u}^h_j(t) = \widehat{u}^h_j(0)e^{-\lambda_j t}$ grows exponentially when $t \to \infty$.

From the Fourier decomposition using the eigenprojections, we can conclude that the L^2-norm of a solution $u(t,x)$ of (6) is $\|u(t,\cdot)\|^2_{L^2([0,\pi])} = \sum_{i=0}^{\infty}(\widehat{u}_i(t))^2$. Since $\lambda_j < 0$ for at least $j = 0$, if we choose an initial condition $\widehat{u}^h_0(0) \neq 0$, then $\widehat{u}_0(t) \to \infty$ when $t \to \infty$. Then, the corresponding solution $u(t,x)$ for such initial condition exhibits grow-up behavior. □

4 Conclusion

Our analysis will enable a future investigation of the asymptotic behavior of this class of equations. Once the existence of the associated non-compact global attractor is ensured, it will be essential to introduce certain objects at infinity that will act as equilibria at infinity. These objects can be characterized with the help of our results and the Fučík spectrum.

Acknowledgments The authors wish to thank Phillipo Lappicy for the extensive and productive discussions on the topic addressed here, which we continue to explore jointly.

References

1. Ben-Gal, N.: Grow-up solutions and heteroclinics to infinity for scalar parabolic PDEs. PhD Thesis. Brown University (2010)
2. Dancer, E.N.: On the Dirichlet problem for weakly non-linear elliptic partial differential equations. Proc. R. Soc. Edinb. Sect. A Math. **76**(4), 283–300 (1977)
3. Fučík, S.: Boundary value problems with jumping nonlinearities. Časopis pro pěstování Mat. **101**(1), 69–87 (1976)
4. Sadyrbaev, F., Gritsans, A.: Nonlinear spectra for parameter dependent ordinary differential equations. Nonlinear Anal. Model. Control **12**(2), 253–267 (2007)
5. Rocha, C., Pimentel, J.: A permutation related to non-compact global attractors for slowly non-dissipative systems. J. Dyn. Differ. Equ. **28**(1), 1–28 (2016)
6. Lappicy, P., Fernandes, J.: Unbounded Sturm attractors for quasilinear parabolic equations. Proc. Edin. Math. Soc. **67**(2), 542–565 (2024)
7. Chepyzhov, V., Goritskii, A.: Unbounded attractors of evolution equations. In: Babin, A.V., Vishik, M.I. (eds.) Properties of Global Attractors of Partial Differential Equations. Adv. Soviet Math., vol. 10, pp. 85–128. American Mathematical Society, Providence (1992)

8. Bortolan, M., Fernandes, J.: Sufficient conditions for the existence and uniqueness of maximal attractors for autonomous and nonautonomous dynamical systems. J. Dyn. Differ. Equ. **1**, 1–30 (2022)
9. Henry, D.: Introduction. In: Geometric Theory of Semilinear Parabolic Equations. Lecture Notes in Mathematics, vol. 840. Springer, Berlin, Heidelberg (1981)
10. Rynne, B.P.: The Fučík Spectrum of General Sturm–Liouville Problems. J. Differ. Equ. **161**(1), 87–109 (2000)

A Computational Framework for the Calculation of a Polynomial Invariant in Spatial Graphs

Myrian Sadith González Orellana and José Mauricio Alvarenga Rodríguez

Abstract Knots have been influential in various research fields, particularly in chemistry where properties of molecules, such as proteins, can be studied using special types of knots known as spatial graphs. This work presents a computational development aimed at creating a software program that automates the calculation of a topological invariant, known as the Yamada polynomial. Different examples of the calculation of the polynomial invariant for different types of molecules will also be found, and aspects such as chirality or the distinction of some proteins will be determined through them. For this work, the language Wolfram Mathematica has been used as the primary tool for computing the required algebra in this work.

Keywords Spatial graphs · Yamada polynomial · Knot theory · Computational framework

1 Introduction

The formal study of knots began with Peter Tait's works in the Proceedings of the Royal Society of Edinburgh in the 1870s. This led to the development of topological invariants, including notable ones such as the Alexander polynomial invariant [2] (1928), the Jones polynomial [6] (1984), and the HOMFLYPT polynomial [5] (1997).

M. S. González Orellana (✉)
School of Mathematics and Computer Science, Universidad Nacional Autónoma de Honduras, Tegucigalpa, Honduras
e-mail: myrian.gonzalez@unah.edu.hn

J. M. Alvarenga Rodríguez
Master's Program in Mathematics, Universidad Nacional Autónoma de Honduras, Tegucigalpa, Honduras
e-mail: jose.alvarenga@unah.edu.hn

P. E. Harris et al. (eds.), *Women in Mathematics in Latin America*,
Research Perspectives Ghent Analysis and PDE Center 14,
https://doi.org/10.1007/978-3-031-97857-9_9

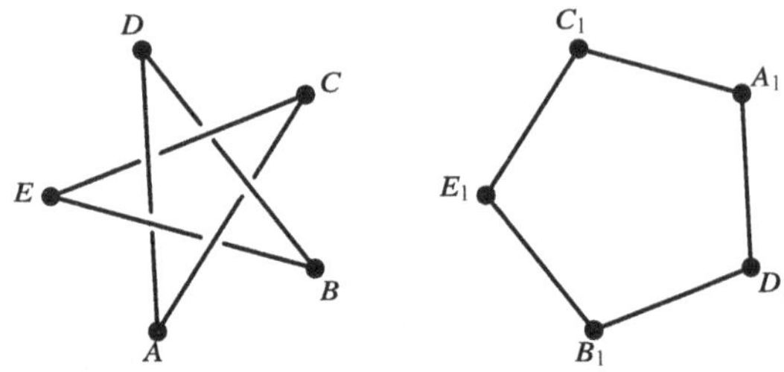

Fig. 1 Both figures can represent isomorphic graphs; however, as spatial graphs, they differ when considering their placement in three-dimensional space [8]

Simultaneously, graph theory applied to chemistry advanced through the works of Harry Wiener [15] (1947) and Milan Randíc [10] (1975), among others.

The intersection of graph theory and knot theory arises naturally since chemical structures exist in three-dimensional space; this can produce non-planar graphs, causing the molecule or structure to lose vital topological information (Graph Theory does not distinguish between the graphs in Fig. 1, despite their differences, as the graph on the left has unavoidable crossings while the other does not). This was notably addressed in Sumners' significant work [12], and this mathematical structure is referred to as a spatial graph [8]. Due to the fact that a spatial graph is not a knot, polynomial invariants such as the Alexander polynomial or the Jones polynomial cannot be directly applied. Significant modifications have been made in this regard, leading to the emergence of polynomial invariants for spatial graphs; one of the most notable and widely used invariants is the so-called Yamada polynomial, introduced in 1989 [17].

Different applications of spatial graph theory to chemistry can be found; among some interesting works are: polymer knot analysis [11] in 2015, protein analysis [4] in 2019.

In the field of computational frameworks for calculating invariant polynomials, works such as Peddada's [9] in 2021, oriented towards the design of an engineering system to address the problem, and Wang's [14] in 2010, where a discussion on a topological quantum computing approach is presented.

Specifically, while various symbolic algebra systems exist, we found no libraries dedicated to this invariant (Yamada polynomial) after thorough searching (though they may exist). Languages like SnapPy or SymPy might allow its construction. Notably, the LinKnot project computes the Jones polynomial and other tools, but no function for the Yamada polynomial was identified.

2 Polynomial Invariant of Spatial Graphs

In this section, a minimal set of definitions will be provided that will lead to the final definition of a spatial graph. Accompanying the definitions, several results will be provided that will be used to support the computational framework. Many of the definitions in this section are taken from sources such as [12] and [8].

First, a spatial graph will be defined from a formal point of view.

Definition 1 Let G be a graph. A spatial graph is the image of a homeomorphism $f\colon G \longrightarrow \mathbb{R}^3$, where the topology of the domain is that of G as a simplicial complex.

Figure 1 illustrates how the spatial graph on the left can be seen as the image of a homeomorphism of the graph on the right. In this sense, it is imagined that the spatial graph exists in three-dimensional space.

Definition 2 Let G be a graph without loops and without multiple edges. Define $\phi_n(G)$ as the number of the way of vertices from G can be colored with n colors, so that two vertices connected by an edge have different colors. Furthermore, define $\rho_n(G) = \phi_n(G)/n$.

A proof of next theorem can be found in reference [3] in the section 7.4.

Theorem 1 *Let G be a graph and e one of its edges. Let G_e and G/e be two graphs formed from G, the first one is G with the edge e removed, and the second one is G with the edge e contracted to a single vertex. Then*

$$\rho_n(G) = \rho_n(G_e) - \rho_n(G/e).$$

Theorem 2 *Suppose G is a planar graph and H_1 and H_2 are two planar representations, then $\rho_n(\overline{H}_1) = \rho_n(\overline{H}_2)$, where $\overline{H}$ represents the dual graph of H.*

To prove the previous theorem, it suffices to note that contracting an edge in the dual is equivalent to deleting an edge in the embedded planar graph (this fact can be utilized by induction on the number of faces). Similarly, deleting an edge in the dual corresponds to contracting an edge in the embedded graph. By performing these operations and leveraging the Theorem 1, we can apply these transformations to both embeddings without altering the isomorphism of the results. Ultimately, by contracting edges, we will obtain a reduced graph with an isomorphic dual, regardless of its embedding in the plane.

To define the Yamada polynomial, the diagram in Fig. 2 is needed. The procedure is described as follows: Let G be a spatial graph. For each crossing of the spatial graph, using the diagram in Fig. 2, three new spatial graphs will be created (if

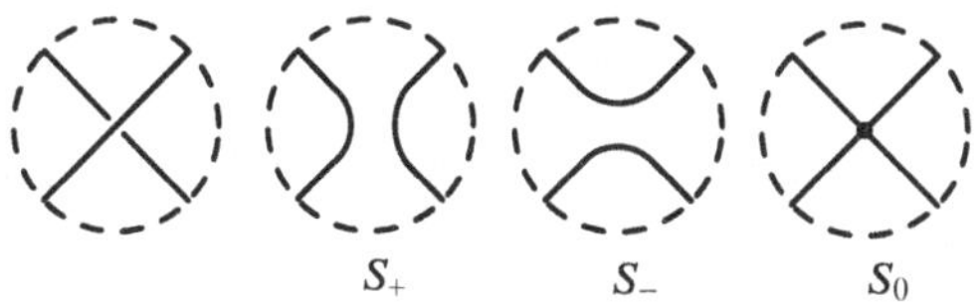

Fig. 2 The image on the left focuses on a crossing of the spatial graph, subsequently, three states are derived, one positive, one negative, and one neutral from left to right denoted by S_1, S_2, and S_3 respectively

there are m crossings, there will ultimately be 3^m resulting planar graphs from the process). Let S_p denote any final state of the process where $p \in \{+, -, 0\}^m$. The Yamada polynomial is then defined as:

$$P(G, t) = \sum_{p \in \{+,-,0\}^m} (-1)^{p_0} t^{p_+ - p_-} \rho_{t+2+t^{-1}}(\overline{S}_p), \tag{1}$$

where in expression (1), p_+, p_-, and p_0 represent the number of positive, negative, and zero signs in p, respectively. Additionally, $\overline{S}_p$ represents the dual graph of S_p, and $\rho_{t+2+t^{-1}}(\overline{S}_p)$ represents the function given in Definition 2 by substituting n with $t + 2 + t^{-1}$.

Another way to define the Yamada polynomial can be achieved with the following recursive formula: assuming that the symbols in parentheses represent a localized crossing of the spatial graph G, it would be necessary to

$$P(\diagdown\!\!\!\!\diagup, t) = t P()(, t) + t^{-1} P(\asymp, t) - P(\times, t), \tag{2}$$

In the previous recursive equation, when reaching the state of the graph where there are no crossings left (if there are m crossings, there will be 3^m terms derived from the recursive equation), we calculate $\rho_n(\cdot)$ and substitute n with $t + t^{-1} + 2$. For more details, refer to the following [8].

To conclude this section, two results found in [17] and [8] related to the Yamada polynomial are presented.

Theorem 3 *Suppose G_1 and G_2 are two equivalent 3-regular spatial graphs, then $P(G_1, t) = \pm t^k P(G_2, t)$ for some integer k.*

Theorem 4 *If G^* is the mirror image of G then $P(G^*, t) = P(G, t^{-1})$.*

3 Computational Framework

In this section, we will begin with a brief explanation of the main functions to be used from Wolfram Mathematica, and then how the spatial graph is provided to the computer program will be discussed. Finally, the implementation of the invariant calculation will be discussed.

Below are the main functions used in this work:

- $G =$Graph[$\cdots$]: This function allows the construction of a graph object. There is a way to provide it with the list of edges of the desired graph as follows:

$$G = \text{Graph}[\{v_{11} \bullet\!\!-\!\!\bullet v_{12}, v_{21} \bullet\!\!-\!\!\bullet v_{22}, \cdots, v_{n1} \bullet\!\!-\!\!\bullet v_{n2}\}]$$

- $D =$DualPlanarGraph[G]: This function generates the dual graph of a planar graph. However, it is important to note that if G is disconnected, then Mathemat-

Fig. 3 Diagram of the states of the input graph for the algorithm

Fig. 4 Types of crossings

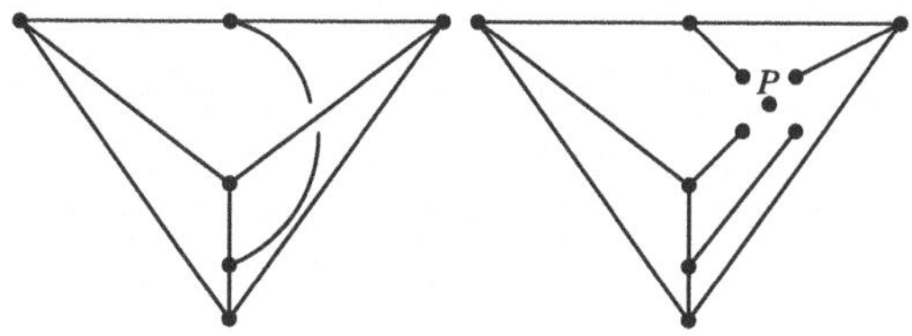

Fig. 5 This spatial graph is an example taken from [8]. The image on the right is the input representation for the program in Wolfram Mathematica; with the encoded information of the crossing type at vertex P

ica returns a disconnected planar graph; this is crucial because not considering it would result in an incorrect calculation of the Yamada polynomial. For more details, see [16].

The input for the Yamada polynomial calculation algorithm will be a graph in the usual sense; however, its vertices will be directly ordered pairs from the Cartesian plane, and for each crossing, five vertices located in squares of length one will be added to the original graph; refer to the first two diagrams (of the five) in Fig. 3. In addition to the graph information, it will be necessary to know if the crossing is of one of the two types shown in Fig. 4 (if we always orient the crossings in the spatial graph projection in the shape of an X). For this purpose, the middle vertex (see the reference in Fig. 3) has been assigned a value of 1 if it is like the one on the left and 0 if it is like the one on the right in Fig. 4.

With the input information, it will be possible to determine which of the last three states on the right in Fig. 3 will be associated with the term t, -1 or t^{-1} in the calculation of the Yamada polynomial. In Fig. 5, we can find the input configuration for the spatial graph on the left.

As explained in the details of the DualPlanarGraph function of Mathematica, the way it operates generates a dual graph for each of its connected components; although, in theory, there should be only one dual graph. To make the correction, we will show that it is possible to recover the dual graph in terms of the disconnected parts generated by such a function in Mathematica: in general, if the resulting graph from performing the operations in the Yamada polynomial calculation process leads to one that has m connected components, then $m-1$ connected components will have their dual in the form of a simple path (this is due to the way the states for the calculation of the Yamada polynomial are generated), that is, they will adopt the form observed in Fig. 6. Next, it will be justified that the modification to be made

Fig. 6 Structure of the connected components of a disconnected state

is simply to divide the contribution of the chromatic function by n^{m-1}. First, we observe that the original dual planar graph is simply the union of these connected parts, and the way this is done is independent and will not affect the final result. To prove the above, we will use the following notation:

- $G + \bullet\!-\!\bullet\!-\!\bullet\cdots\bullet\!-\!\bullet$ is the graph obtained by identifying exactly one of the vertices of the simple path with any of the vertices of the graph G.
- $G \oplus \bullet\!-\!\bullet\!-\!\bullet\cdots\bullet\!-\!\bullet$ is the graph obtained by adding the simple path to G in a disconnected way.

First, we will prove that adding an edge is equivalent to multiplying the original graph by $(n-1)$:

$$\rho_n(G + \bullet\!-\!\bullet) = \rho_n(G \oplus \bullet) - \rho_n(G) = n\rho_n(G) - \rho_n(G) = (n-1)\rho_n(G).$$

As a second point, consider the following relationships involving a simple path with k edges:

$$\rho_n(G + \bullet\!-\!\bullet\!-\!\bullet\cdots\bullet\!-\!\bullet) = (n-1)^k \rho_n(G).$$

$$\rho_n(G \oplus \bullet\!-\!\bullet\!-\!\bullet\cdots\bullet\!-\!\bullet) = (n-1)^{k-1} \rho_n(G \oplus \bullet\!-\!\bullet) = (n-1)^k n\rho_n(G).$$

As can be seen from the above, it follows that $\rho_n(G + \bullet\!-\!\bullet\!-\!\bullet\cdots\bullet\!-\!\bullet) = \dfrac{\rho_n(G \oplus \bullet\!-\!\bullet\!-\!\bullet\cdots\bullet\!-\!\bullet)}{n}$.

3.1 Algorithm Implementation

Initially, three structures will be needed for the use of this function: the coordinates of the vertices of the graph, the construction of the graph, and an encoding for the type of crossings:

Vertex Structure, Graph Construction

The following code defines a structure that will then be used to adjust the coordinates of each vertex. The graph will be constructed directly from the edges as shown in the code. For the encoding of the crossings, assume that the first n vertices correspond to these crossings. Then, 1 or 0 will be assigned

(continued)

to these crossings, based on whether they are of the type on the left or right as shown in Fig. 4, respectively.

```
1 V1={x1,y1};V2={x2,y2};··· Vr={xr,yr};
2 Coor={V1-> V1,···,Vr-> Vr};
3 Edge={Vi1 •-•Vj1, ··· , Vis •-•Vjs };
4 G=Graph[Edge,VertexCoordinates -> Coor];
5 Cross={{V1,l1},{V2,l2},···,{Vn,ln}};
```

In line 5, l_i represents one of the two types of crossings, encoded as 0 or 1.

Function for Calculating the Yamada Polynomial

Some further important aspects of the developed function will be described: in line 2, *Module* was used so that the variables in the main environment do not interfere with the variables used in the function and vice versa.

```
1  Pol[G_,cross_,Coor_]:=\
2  Module[{a,Connect,DGA,factor,GA,j1,k,k1,list,m,n,Poly,state,x,x1,y1},
3  Poly=0; n=Length[cross]; list =Tuples[{0,1,-1},n];k=Length[list];
4  For[j1=1,j1<=k,j1++,
5      state=list[[j1]];GA=G;
6      For[k1=1,k1<=n,k1++,
7          x1=cross[[k1]][[1]][[1]];y1=cross[[k1]][[1]][[2]];
8          If[state[[k1]]==0,
9              (GA=EdgeAdd[GA,{{x1,y1}•-•{x1-1,y1-1},{x1,y1}•-•{x1-1,y1+1}}];
10             GA=EdgeAdd[GA,{{x1,y1}•-•{x1+1,y1-1},{x1,y1}•-•{x1+1,y1+1}}];),
11             (a=state[[k1]]*(1-2*cross[[k1]][[2]]);
12             GA=EdgeAdd[GA,{{x1-1,y1+1} •-• {x1+a,y1+a},{x1-a,y1-a} •-• {x1+1,y1
                   -1}}];
13             GA=VertexDelete[GA,{x1,y1}];)];];
14         DGA=SimpleGraph[DualPlanarGraph[GA]];
15         Connect=ConnectedGraphComponents[DGA];
16         m=Length[Connect];factor=(-1)^Count[state,0]*t^Total[state]/x^m;
17         Poly=Poly+factor*ChromaticPolynomial[DGA,x];];
18         x=t+t^(-1)+2;
19         Poly];
```

In line 3, a list is created with all possible states defined in the crossings. Each state is a tuple where its values can be −1, 0, or 1; this corresponds to one of the last three diagrams in Fig. 3 depending on the type of crossing.

Between lines 6 and 13, the corresponding graph for that state is constructed. From lines 14 to 17, the contribution to the Yamada polynomial from this state is calculated. In line 16, it can be seen how it is divided by the term x^m (previously denoted by n^m) as anticipated.

4 Case Studies

Next, the implementation will be tested for calculating the invariant in different cases.

Case 1 Consider initially the example found in [8] shown the image on the left in Figure 5 within this article. If we apply our code, we find that:

$$P(G,t) = -2t^3 + t^2 - 4t + 3 - 3t^{-1} + 4t^{-2} - t^{-3} + 2t^{-4}.$$

As one can verify, this spatial graph is equivalent to what would be seen in a mirror. It is observed that $P(G,t^{-1}) = -tP(G,t)$, and thus in this case, the invariant could not distinguish if the graph is chiral.

Case 2 Another example taken from [8] is the type of spatial graph M_4 (more generally M_n); this spatial graph arises from the generalization of M_3 representing the molecule found by D.M. Walba in [13]. Using the implementation, we obtain:

$$t^7P(G,t) = -2t^{12} + 4t^{11} - 16t^{10} + 19t^9 - 37t^8 + 35t^7$$
$$-44t^6 + 36t^5 - 31t^4 + 20t^3 - 11t^2 + 4t - 1$$

and

$$t^5P(G,t^{-1}) = -t^{12} + 4t^{11} - 11t^{10} + 20t^9 - 31t^8 + 36t^7$$
$$-44t^6 + 35t^5 - 37t^4 + 19t^3 - 16t^2 + 4t - 2.$$

As can be seen, their invariants are different, and therefore, the spatial graph is chiral.

Case 3 Let's analyze a chemical derivative found in [7]; this can be seen in Fig. 7. The output is shown below:

$$t^8P(G,t) = t^{14} + 5t^{13} - 17t^{12} + 46t^{11} - 75t^{10} + 110t^9 - 125t^8 + 134t^7$$
$$-116t^6 + 94t^5 - 57t^4 + 30t^3 - 8t^2 + t + 1$$

and

$$t^6P(G,t^{-1}) = t^{14} + t^{13} - 8t^{12} + 30t^{11} - 57t^{10} + 94t^9 - 116t^8 + 134t^7$$
$$-125t^6 + 110t^5 - 75t^4 + 46t^3 - 17t^2 + 5t + 1.$$

Fig. 7 The graph represents a part of a molecule found in [7]

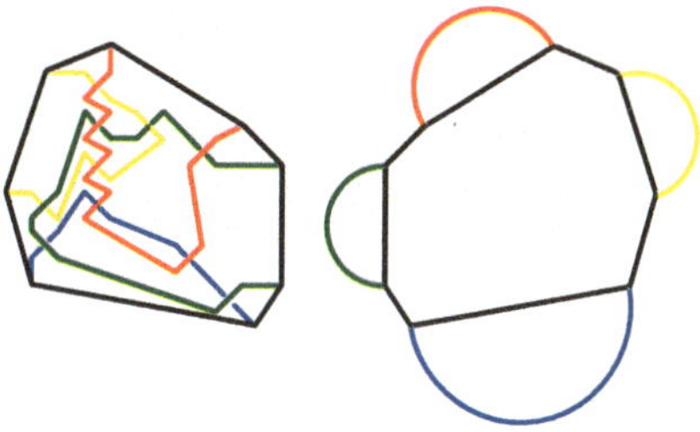

Fig. 8 Ten-crossing graph. Both spatial graphs are equivalent

As in the previous case, we have the chirality property for any compound that adopts this form.

Case 4 To conclude this work, a graph with ten crossings was chosen to validate the code's capability. In this case, the graph shown in Fig. 8 was used for this purpose. The left side of the figure shows the spatial graph with 10 crossings, while the right side displays its equivalent without crossings. It is evident how by moving the red, green, yellow, and blue parts, in that order, we can obtain the graph on the right. After running the code: $P(G,t) = \frac{(1+t^2)^4(1+t+t^2)}{t^5}$. This is the same term associated with the chromatic polynomial of the right figure.

5 Conclusions

This article has demonstrated the construction of a computational framework that ultimately allowed the development of a function for calculating the Yamada polynomial using a simple access program like Wolfram Mathematica. Also, this work allows for easy access to code usage and replication.

Different cases have been presented that validate and support the correct operation of the entire computational framework. Particularly, two application cases directly related to two types of chiral compounds found in different articles were presented.

There are possible improvements that can be made to the work presented here. Regarding runtime performance, exploring the entire state space (of order 3^m where m is the number of crossings) is difficult; however, a spatial graph with 10 crossings could be processed, which, although not an exorbitant amount, can be applied to many cases that would be a challenging calculation without computer use. To improve runtime, two future work proposals are suggested: firstly, making use of parallel programming paradigm since the proposed algorithm is easily parallelizable. Secondly, exploit the properties of the Yamada polynomial, which would allow the reduction of the number of state spaces considerably.

This article has discussed the Yamada polynomial. However, there are other closely related invariants that can be explored based on the calculation method used in this work. For instance, consider the research conducted in [1] which exposes an invariant of spatial graphs.

References

1. Adams, C., Devadoss, J., Elhamdadi, M., Mashaghi, A.: Knot theory for proteins: Gauss codes, quandles and bondles. J. Math. Chem. **58**, 1711–1736 (2020)
2. Alexander, J.W.: Topological invariants of knots and links. Trans. Am. Math. Soc. **30**(2), 275–306 (1928)
3. Bondy, J.A., Murty, U.S.R.: Graph Theory. Springer, London (2008)
4. Flapan, E., Heller, G.: Topological complexity in protein structures. Comput. Math. Biophys. **3**(1), 23 (2015)
5. Freyd, P., Yetter, D., Hoste, J., Lickorish, W.R., Millett, K., Ocneanu, A.: A new polynomial invariant of knots and links. Bull. Am. Math. Soc. **12**(2), 239–246 (1985)
6. Jones, V.F.: A polynomial invariant for knots via von Neumann algebras. In: Fields Medallists' Lectures, pp. 448–458 (1997)
7. Liang, C., Mislow, K.: Topological chirality of proteins. J. Am. Chem. Soc. **116**(8), 3588–3592 (1994)
8. Murasugi, K., Kurpita, B.: Knot Theory and Its Applications, pp. 1–341. Birkhäuser, Boston (1996)
9. Peddada, S.R., Dunfield, N.M., Zeidner, L.E., Givans, Z.R., James, K.A., Allison, J.T.: Enumeration and identification of unique 3D spatial topologies of interconnected engineering systems using spatial graphs. J. Mech. Des. **145**(10), 101708 (2023)
10. Randic, M.: Characterization of molecular branching. J. Am. Chem. Soc. **97**(23), 6609–6615 (1975)
11. Shimokawa, K., Ishihara, K., Tezuka, Y.: Knot theory analysis of polymers. In: Topology of Polymers. SpringerBriefs in the Mathematics of Materials, vol. 4. Springer, Tokyo (2019)
12. Sumners, D.W.: The knot theory of molecules. J. Math. Chem. **1**(1), 1–14 (1987)
13. Walba, D.M., Richards, R.M., Haltiwanger, R.C.: Total synthesis of the first molecular Möbius strip. J. Am. Chem. Soc. **104**(11), 3219–3221 (1982)
14. Wang, Z.: Topological Quantum Computation, No. 112. American Mathematical Society, Providence (2010)
15. Wiener, H.: Structural determination of paraffin boiling points. J. Am. Chem. Soc. **69**(1), 17–20 (1947)
16. Wolfram Research: DualPlanarGraph, Wolfram Language function (2021). https://reference.wolfram.com/language/ref/DualPlanarGraph.html
17. Yamada, S.: An invariant of spatial graphs. J. Graph Theory **13**(5), 537–551 (1989)

The Search for Rainbow 3-Term Arithmetic Progressions: An Inspirational Journey

Amanda Montejano

Abstract We summarize known results on the existence of rainbow 3-term arithmetic progressions within 3-colorings of various ground sets, such as finite intervals, cyclic groups, and general Abelian groups. This problem represents the foundational question in the study of the existence of rainbow structures under density conditions on the color classes, inspiring numerous works and generalizations within this area of Rainbow Ramsey Theory.

Keywords Rainbow arithmetic progressions · Ramsey theory · Rainbow structures

1 Introduction

Ramsey Theory stands as one of the most captivating areas in Combinatorics. The intuitive nature of many of its results and open problems, combined with the elegance of its proofs and the diversity of the methods employed, make it a jewel in the field of mathematics. What started as a collection of seemingly disconnected theorems, all sharing the philosophy that complete disorder is impossible, has become a cohesive and influential branch of contemporary mathematics. The charm of Ramsey Theory is also enriched by the wide range of ideas and techniques it incorporates from different areas, along with the fascinating variations that have emerged within the theory. Here, we seek to illustrate the points made above.

Although not all results—particularly the older ones—were established using the language of colorings, employing such a framework is often very useful for

This work was supported by UNAM PASPA-DGAPA.

A. Montejano (✉)
UMDI, Facultad de Ciencias, UNAM Juriquilla, Juriquilla, Querétaro, Mexico
e-mail: amandamontejano@ciencias.unam.mx

P. E. Harris et al. (eds.), *Women in Mathematics in Latin America*,
Research Perspectives Ghent Analysis and PDE Center 14,
https://doi.org/10.1007/978-3-031-97857-9_10

presenting the results. Given a set S and a positive integer r, any mapping from S to a set of cardinality r (whose elements are called colors) is said to be an *r-coloring* of S. A subset $T \subseteq S$ is called *monochromatic* with respect to a given coloring if the mapping assigns the same color to every element in T. Arithmetic progressions are one of the key objects of study in this area. A set of k distinct integers $\{a, a+d, \cdots, a+(k-1)d\}$ is called a *k-term arithmetic progression* $AP(k)$. Using only these definitions, we can state one of the most fundamental theorems in Ramsey Theory: van der Waerden's theorem from 1927, which asserts that any r-coloring of the natural numbers contains a monochromatic arithmetic progression of any finite length k. There are many excellent introductory texts on Ramsey Theory, including the classic book written by Ronald Graham, Bruce Rothschild and Joel Spencer [11] from which many have learned, as well as the more recent books by Landman and Roberson [16], and Jungic [13].

In this paper, we focus on *Rainbow Ramsey Theory*, also known as *anti–Ramsey Theory*, where the aim is to study the existence of rainbow subsets within colorings of large sets. Formally, a subset $T \subseteq S$ is called *rainbow* with respect to a coloring c, if $c(x) \neq c(y)$ for all $x, y \in T$, $x \neq y$ (meaning that c assigns pairwise distinct colors to the elements of T). There are several approaches in the study of the existence of rainbow structures in colored mathematical universes. For instance, the *canonical Ramsey Theory* initiated in 1950 with a combinatorial theorem by Erdős and Rado [7] identify three fundamental types of homogeneous patterns: monochromatic, rainbow, and a third option (a kind of lexicographic color pattern) which is less simple to describe. There are situations, however, mostly in the arithmetic context, where the canonical versions of classical results lack the third option. For example, the canonical version of van der Waerden's theorem, due to Erdős and Graham [6], asserts that any coloring of the natural numbers (possibly with infinitely many colors) contains either a monochromatic or a rainbow $AP(k)$ for any positive integer k.

To ensure the existence of rainbow subsets independent of monochromatic ones, it is necessary to have a sufficient number of colors and to guarantee that each color is sufficiently well represented; otherwise, a constant coloring would never yield a rainbow set. There are two natural ways to address this situation: either by imposing a bound on the number of times that a color *can* be used (this interesting approach is beyond the scope of this text; we referred the reader to [1, 4] for more information), or by imposing a bound on the number of times that a color *must* be used. Such a condition can be as simple as requiring the presence of each color. In graph theory, the problem of the existence of rainbow subsets with such a minimal condition has been extensively studied, particularly regarding the minimum number of colors required to ensure that every surjective coloring (a coloring that utilizes all available colors) produces a rainbow structure. The origin of this parameter is attributed to Erdős, Simonovits, and Sós in 1975 [8], where they defined the following integer-valued function: given a positive integer n and a graph G, let $f(n, G)$ denote the maximum number of colors such that the set of edges of a complete graph on n vertices, K_n, can be colored without producing a rainbow copy of G. In this paper, however, we are interested in a more general and recent approach, introduced in

2003 by Jungić et al. [14] in the context of Arithmetic Ramsey Theory. The novelty of their perspective lies in fixing the number of colors and studying the existence of rainbow subsets under specific density constraints on the coloring.

Given a coloring $c : S \to \{1, \cdots, r\}$, the sets $c^{-1}(i) \subseteq S$, for $1 \leq i \leq r$, are called the *color classes* of c. It should be noted that, a coloring c can be thought as a partition of S into the color classes, $S = c^{-1}(1) \cup c^{-1}(2) \cup \cdots \cup c^{-1}(r)$. The study of the existence of rainbow structures—independently of monochromatic ones—by analyzing the size or density of the color classes, becomes particularly relevant in view of Szemerédi's theorem, a very powerful result that often regarded as the "density version" of van der Waerdens Theorem. Roughly speaking, Szemerédi's theorem asserts that, for any positive integer k, any subset of the natural numbers with positive upper density contains a k-term arithmetic progression $AP(k)$ [20]. Therefore, any coloring with a sufficiently large color class will contain a monochromatic arithmetic progression. What would an anti-Ramsey version of van der Waerden's theorem look like? The problem is far from trivial, given the existence of colorings of $\mathbb{N}$ with infinitely many colors, each having positive density, yet containing no rainbow $AP(3)$ (for an example of such a coloring, see [14]).

The earliest results in the literature regarded as rainbow counterparts of classical results in Arithmetic Ramsey Theory, deal with the existence of rainbow 3-term arithmetic progressions in every 3-coloring with sufficiently many elements of each color. Essentially, while Szemerédi's theorem guarantees the existence of a monochromatic arithmetic progression (of any given length) within a single "large" color class, we need all color classes to be "large" in order to guarantee a rainbow $AP(3)$. This phenomenon will be formally examined throughout the rest of the paper by exploring the existence of rainbow 3-term arithmetic progressions in 3-colorings of various ground sets, such as the set of natural numbers $\mathbb{N}$ (Sect. 2.1), finite integer intervals (Sect. 2.2), cyclic groups (Sect. 2.3) and general Abelian groups (Sect. 2.4). To conclude, in Sect. 3, we highlight related research in the literature that has been inspired by the search for rainbow 3-term arithmetic progressions.

2 Rainbow 3-Term Arithmetic Progressions

Let $n \geq 3$ be an integer. It is easy to see that not every 3-coloring of the integer interval $[1, n]$ contains a rainbow 3-term arithmetic progression. This remains true (for $n \geq 4$) even when considering colorings that use all colors, that is, colorings with non-empty color classes. A 3-coloring $c : [1, n] \to \{r, b, g\}$ with all color classes, $c^{-1}(r) = R$, $c^{-1}(b) = B$ and $c^{-1}(g) = G$, of the same size, $|R| = |B| = |G|$, is called an *equinumerous* coloring. According to [14], during a problem session at MIT in 2001, Radoš Radoičić posed the question of whether, for every $n \equiv 0 \pmod 3$, any equinumerous 3-coloring of $[1, n]$ must contain a rainbow $AP(3)$. And so, this line of inquiry began.

2.1 *The Infinite Version*

Before Radoičić's question was answered positively, an infinite version of the problem had been established. The *upper density* of a set $S \subset \mathbb{N}$ is defined as $\limsup_{n\to\infty} \frac{|S\cap[1,n]|}{n}$. Recall that it was pointed out in the introduction that there are colorings of $\mathbb{N}$ with infinitely many colors, each with positive upper density, yet containing no rainbow $AP(3)$ [14] (see also [15]). This leads to the natural question: what upper density, if any, on the color classes would force the existence of a rainbow $AP(3)$? The answer is given in the next result.

Theorem 1 (Jungić et al. [14]) *Every* 3*-coloring of* $\mathbb{N}$ *with the upper density of each color class greater than* $1/6$ *contains a rainbow* $AP(3)$.

A coloring that contains no rainbow $AP(3)$ is said to be $AP(3)$*-rainbow-free*. Note that the coloring defined as

$$c(i) = \begin{cases} r & \text{if } i \equiv 1(\text{mod}6), \\ b & \text{if } i \equiv 4(\text{mod}6), \\ g & \text{otherwise.} \end{cases} \tag{1}$$

is $AP(3)$-rainbow-free with the upper density of its color classes being $1/6$, $1/6$, and $2/3$, respectively. Thus, the fraction $1/6$ in Theorem 1 is best possible.

Based on the coloring given in (1) and computer evidence, the authors of [14] propose a conjecture regarding the maximum size that the smallest color class in an $AP(3)$-rainbow-free 3-coloring of $[1, n]$ can have. This conjecture, known as the finite version of Theorem 1, was confirmed by Axenovich and Fon-Der-Flaass [3].

2.2 *The Finite Version*

For any positive integer n, let us examine the restriction of the coloring defined in (1) to the interval $[1, n]$. For example, if $n \leq 24$, see the coloring depicted below:

1 2 3 4 5 6 7 8 9 10 11 12 13 14 15 16 17 18 19 20 21 22 23 24.

Note that the smallest color class of this coloring is always $\lfloor \frac{n+2}{6} \rfloor$. Define $\mu(n)$ to be the largest cardinality of the smallest color class over all $AP(3)$-rainbow-free 3-colorings of $[1, n]$. Coloring (1) implies $\mu(n) \geq \lfloor \frac{n+2}{6} \rfloor$. The exact value of $\mu(n)$ was established in [3]. It turns out that $\mu(n) = f(n)$, where $f(n)$ is the following integer-valued function:

$$f(n) = \begin{cases} \frac{n+4}{6} & \text{if } n \equiv 2(\text{mod } 6), \\ \lfloor \frac{n+2}{6} \rfloor & \text{otherwise.} \end{cases} \tag{2}$$

In other words, for $n \not\equiv 2 \pmod 6$ the best we can hope for a rainbow-free 3-coloring of $[1, n]$ is to have the smallest color class as large as the one given by (1). But, for $n \equiv 2 \pmod 6$ there is a 3-coloring with no rainbow 3-term arithmetic progressions with slightly larger cardinality of the smaller color class. Besides that, the hard part in proving $\mu(n) = f(n)$ is to show that every coloring of $[1, n]$ with all three color classes greater than $f(n)$ contain a rainbow $AP(3)$. This was conjectured in [14] and proved in [3].

Theorem 2 (Axenovich, Fon-Der-Flaass [3]) *For every integer $n \geq 3$, every 3-coloring $[1, n] = R \cup B \cup G$ with $\min\{|R|, |B|, |G|\} > f(n)$ contains a rainbow $AP(3)$.*

The proofs of Theorems 2 and 1 rely on the fact that every rainbow-free coloring of $[1, n]$ or $\mathbb{N}$ contains a special color known as the "dominant" color. In both cases, the strategy is to argue that if the smallest color class is not sufficiently large (or dense) then it can't be a dominant color. The absence of the dominant color leads to the existence of a rainbow $AP(3)$. In summary, there is an intriguing connection between the density of the smallest color class, the largest (dominant) color, and the presence of a rainbow $AP(3)$. When the set to be colored is an Abelian group, we have tools that offer deeper insights into this behavior.

2.3 *The Modular Version*

Observe that an $AP(3)$ in $\mathbb{N}$ is a non-trivial integer solution of the equation $x + y = 2z$. A *non-trivial* solution means not to have $x = y = z$. The modular version of the problem considers non-trivial solutions to $x + y \equiv 2z \pmod n$, which can be interpreted as coloring modular classes instead of integers (the coloring described in (1) exemplifies this). Equivalently, we color the elements of the cyclic group $\mathbb{Z}_n$ and look for a rainbow $AP(3)$ within the group $\mathbb{Z}_n$. Naturally, any arithmetic progression in $[1, n]$ will also be an arithmetic progression in $\mathbb{Z}_n$, but the converse is not necessarily true. For this reason, it follows directly from Theorem 1 that every 3-coloring of $\mathbb{Z}_n$, where the cardinality of each color class is greater than $n/6$, contains a rainbow $AP(3)$. However, in this context, the situation becomes more interesting since the proportion $n/6$, which is tight for every $n \equiv 0 \pmod 6$, by the coloring described in (1), can be improved for most other values of n depending on its smallest prime factors.

As we defined $\mu(n)$ in Sect. 2.2, we now define $\mu(\mathbb{Z}_n)$ as the largest cardinality of the smallest color class over all $AP(3)$-rainbow-free 3-colorings of $\mathbb{Z}_n$. In the following theorem, we summarize the bounds of $\mu(\mathbb{Z}_n)$ that were explored in [14].

Theorem 3 (Jungić et al. [14]) *Let n be a positive integer such that n is not a power of 2, q is the smallest prime factor of n, and p is the smallest odd prime factor of n. Then,*

$$\left\lfloor \frac{n}{2p} \right\rfloor \leq \mu(\mathbb{Z}_n) \leq \min\left\{\frac{n}{6}, \frac{n}{q}\right\}. \tag{3}$$

For example, if $n \equiv 0 \pmod 6$ then $q = 2$ and $p = 3$, thus $\mu(\mathbb{Z}_n) = n/6$. Note that $q \leq p$ with equality if and only if n is odd. Also, the upper bound in (3) remains being $n/6$ when $q \in \{2, 3, 5\}$, but can be arbitrary lowered as q grows. Furthermore, if $n > 5$ is a prime number, then $n = q = p$ and $0 \leq \mu(\mathbb{Z}_n) \leq 1$. This means that for every prime number p, either $\mu(\mathbb{Z}_p) = 0$ or $\mu(\mathbb{Z}_p) = 1$. The positive integers n for which $\mu(\mathbb{Z}_n) = 0$ were fully characterized in [14]. Specifically, it was shown that $\mu(\mathbb{Z}_n) = 0$ if and only if n satisfies one of the following two conditions:

(i) n is a power of 2.
(ii) n is a prime with the multiplicative order of 2 being either $n - 1$, or $(n - 1)/2$ with $(n - 1)/2$ odd.

This gives a partition of the set of odd prime numbers $\mathcal{P}$. Let $\mathcal{P}_0$ be the set of odd primes p for which 2 has either multiplicative order $p - 1$, or multiplicative order $(p - 1)/2$ with $(p - 1)/2$ odd, and let $\mathcal{P}_1 = \mathcal{P} \setminus \mathcal{P}_0$. Thus, for $i \in 0, 1$, we have $\mu(\mathbb{Z}_p) = i$ if and only if $p \in \mathcal{P}_i$. The following result, concerning cyclic groups of order n, generalizes this information. This conjecture was proposed in 2005 [15] and subsequently confirmed in 2012 [19].

Theorem 4 (Conjectured in [15], Proved in [19]) *Let n be an integer which is not a power of 2. Let p be the smallest prime factor of n in $\mathcal{P}_0$ and q be the smallest prime factor of n in $\mathcal{P}_1$. Then, $\mu(\mathbb{Z}_n) = \left\lfloor \frac{n}{\min\{2p,q\}} \right\rfloor$.*

Montejano and Serra proved the result of Theorem 4 not only for cyclic groups but for Abelian groups of odd order.

2.4 Finite Abelian Groups

To begin this section, we point out that Theorem 4 is stated only for cyclic groups but, as mentioned before, it is also true for Abelian groups of odd order. However, the statement does not hold for general Abelian groups of even order. This means that there are Abelian groups of even order that can be colored using three colors in such a way that no rainbow $AP(3)$ exists, and the cardinality of the smallest color class is much larger than what is indicated in Theorem 4. The example given below, provided originally in [19], illustrates this situation. To properly state it, we need some terminology. Let Γ be a finite Abelian group and $H < \Gamma$ be a subgroup of Γ. We denote by $\pi : \Gamma \to \Gamma/H$ the natural projection of Γ into the quotient group

Γ/H. Given a coloring $c : G/H \to \{r, b, g\}$ we define the *lifting coloring*, $c' : \Gamma \to \{r, b, g\}$, as $c'(x) = c(\pi(x))$. Since a triple (x, y, z) in Γ satisfying $x+z = 2z$ maps into a triple $(\pi(x), \pi(y), \pi(z))$ in Γ/H satisfying the same equation, then the lifting coloring of an $AP(3)$-rainbow-free coloring is also rainbow-free.

Example Consider the Klein group $K = \mathbb{Z}_2 \oplus \mathbb{Z}_2$, which is a noncyclic group of order 4 containing no (non-trivial) 3-term arithmetic progressions. Therefore, any 3-coloring of K is $AP(3)$-rainbow-free. Let H be any finite Abelian group and define $\Gamma = H \oplus K$. The lifting coloring of any 3-coloring of K with non-empty color classes, results in an $AP(3)$-rainbow-free coloring of Γ with the smallest color class of cardinality $|H|$. This quantity can be arbitrarily larger than $\left\lfloor \frac{n}{\min\{2p,q\}} \right\rfloor$ by choosing H as large as desired. □

The case of general Abelian groups of odd order behaves differently. As previously outlined, Montejano and Serra confirm the result of Theorem 4 not only for cyclic groups but for Abelian groups of odd order. Most notably, instead of merely determining $\mu(\Gamma)$—the maximum size of the smallest color class among all $AP(3)$-rainbow-free 3-colorings of Γ—they provide an explicit description of every 3-coloring of Γ that avoids rainbow 3-term arithmetic progressions. This description is built from $AP(3)$-rainbow-free 3-colorings of prime groups.

The description of $AP(3)$-rainbow-free colorings for groups of prime order is relatively simple. As discussed in the paragraph before Theorem 4, if $p \in \mathcal{P}_0$, then $\mu(\mathbb{Z}_p) = 0$, meaning there are no rainbow-free 3-colorings with non-empty color classes (equivalently, every $AP(3)$-rainbow-free coloring uses at most two colors). On the other hand, if $p \in \mathcal{P}_1$, then $\mu(\mathbb{Z}_p) = 1$, which indicates that an $AP(3)$-rainbow-free 3-coloring either has an empty color class or, if not, there is a color class with just one element, an advantage for describing rainbow-free colorings. To proceed, we need some more notation. For a subset $X \subset \Gamma$ let $-X = \{-x : x \in X\}$ and $2X = \{2x : x \in X\}$. The next proposition provides a characterization of $AP(3)$-rainbow-free colorings of groups of prime order. Note that the property of being $AP(3)$-rainbow–free is invariant under translation.

Proposition 5 (Montejano, Serra [19]) *Let p be a prime. A coloring $\mathbb{Z}_p = R \cup B \cup G$ with non-empty color classes is $AP(3)$-rainbow-free if and only if, up to translation, there is a color class, say R, such that $R = \{0\}$ and both $X \in \{B, G\}$ satisfy $X = 2X = -X$.*

Proof By Theorem 3, we know that $0 \leq \mu(\mathbb{Z}_p) \leq 1$. Hence, a 3-coloring with non-empty color classes is $AP(3)$-rainbow-free if and only if the smallest color class, say R, has cardinality one. Since the property of being $AP(3)$-rainbow-free is invariant under translation, we can assume that $R = \{0\}$. Now observe that, in $\mathbb{Z}_p$, a 3-term arithmetic progression containing 0 must be of the form $(-x, 0, x)$ or $(0, x, 2x)$. Thus, the coloring will be rainbow-free if and only if $2B = B = -B$ and $2G = G = -G$. □

It is worth mentioning that a partition of $\mathbb{Z}_p \setminus \{0\}$ into non-empty sets B and G satisfying $B = 2B = -B$ and $G = 2G = -G$ exist if and only if $p \in \mathcal{P}_1$. The smallest element of $\mathcal{P}_1$ is $p = 17$, and the following is the only $AP(3)$-rainbow-free coloring up to translation

$$\mathbb{Z}_{17} = \{0\} \cup \{1, 2, 4, 8, 9, 13, 15, 16\} \cup \{3, 5, 6, 7, 10, 11, 12, 14\}.$$

For a general Abelian group Γ, there are two natural ways to construct an $AP(3)$-rainbow-free coloring. The first one consists of selecting, if it exists, a proper subgroup $H < \Gamma$ such that $|\Gamma/H| = p \in \mathcal{P}_1$, and lifting an $AP(3)$-rainbow-free 3-coloring with non-empty color classes from Γ/H to Γ. The second one consists of choosing any proper subgroup $H < \Gamma$, coloring its elements red and blue in any way, and assigning color green to the remaining elements. The main theorem in [19] identifies three possible types of $AP(3)$-rainbow-free 3-colorings of Abelian groups of odd order: either the coloring fits one of the forms described above, or it is a combination of both.

Theorem 6 (Montejano, Serra [19]) *Let Γ be a finite Abelian group of odd order n. A coloring $\Gamma = R \cup B \cup G$ with non-empty color classes is $AP(3)$-rainbow-free if and only if, up to translation, there is a proper subgroup $H < \Gamma$ and a color class, say R, such that the following three conditions hold:*

(i) *$R \subseteq H$, and the 3-coloring induced in H is rainbow-free,*
(ii) *both $\widetilde{B} = B \setminus H$ and $\widetilde{G} = G \setminus H$ are the union of H-cosets, and*
(iii) *$\widetilde{B} = -\widetilde{B} = 2\widetilde{B}$ and $\widetilde{G} = -\widetilde{G} = 2\widetilde{G}$.*

In some sense, up to translation, every rainbow-free coloring of Γ is built by selecting a subgroup H and combining two rainbow-free colorings: one for H and the other for Γ/H, where the rainbow-free coloring of Γ/H has a color class with just one element. In this way, rainbow-free colorings of Abelian groups of odd order are nested compositions of rainbow-free colorings of smaller groups. The fact that the smallest color class is contained in a proper subgroup of Γ implies Theorem 4. To prove Theorem 6, powerful results from additive number theory were used. Such as the theorem of Kneser, Kemperman, and Grynkiewicz which gives the structure of sets with small sumset in Abelian groups (see [19]).

3 Further Research

This brief manuscript focuses on the existence of rainbow 3-term arithmetic progressions in 3-colorings of various ground sets. Numerous related problems have been studied in this area. For example, rainbow-free colorings for general systems of linear equations, as in Rado's Theorem in Ramsey Theory, were explored in [5], while specific equations in three variables were addressed in [12] and [17]. Studies of the Sidon equation, a classic topic in Additive Number Theory, appear in [9] and [10]. Although much is known about $AP(k)$-rainbow-free colorings, intriguing questions remain open (see [18]). Even the impossibility of a rainbow counterpart of the Hales-Jewett theorem (see [4]) yields new insights and inspired beautiful problems (see [2]).

Acknowledgments This chapter was supported by a Postdoctoral fellow from DGAPA UNAM and PAPIIT project IG100822.

References

1. Alon, N., Caro, Y., Tuza, Z.: Sub-Ramsey numbers for arithmetic progressions. Graphs Combin. **5**, 307–314 (1989)
2. Araujo-Pardo, G., Fernández-Merchant, S., Hansberg, A., Lara, D., Montejano, A., Oliveros, D.: The exact balanced upper chromatic number of the n-cube over t-elements (preprint). Short version atn: The 36th Canadian Conference on Computational Geometry (2024)
3. Axenovich, M., Fon-Der-Flaass, D.: On rainbow arithmetic progressions. Electron. J. Combin. **11**, R1 (2004)
4. Axenovich, M., Martin, R.: Sub-Ramsey numbers for arithmetic progressions. Graphs Combin. **22**(3), 297–309 (2006)
5. De Loera, J.A., La Haye, R.N., Montejano, A., Oliveros, D., Roldán, E.: A rainbow Ramsey analogue of Rado's theorem. Discrete Math. **339**, 2812–2818 (2016)
6. Erdős, P., Graham, R.L.: Old and New Problems and Results in Combinatorial Number Theory, vol. 28. L'Enseignement Math, Monographie, Geneva (1980)
7. Erdős, P., Rado, R.: A combinatorial problem. J. Lond. Math. Soc. **25**, 249–255 (1950)
8. Erdős, P., Simonovits, M., Sós, V.: Anti-Ramsey theorems. In: Infinite and Finite Sets, Colloq. Math. Soc. Janos Bolyai, Discrete Math., vol. 10, pp. 633–643 (1975)
9. Espinosa-García, M.A., Montejano, A., Roldán-Pensado, E., Suárez, J.D.: Sidon-Ramsey and B_h-Ramsey numbers. Bol. Soc. Mat. Mex. **30**, 104, 1–14 (2024)
10. Fox, J., Mahdian, M., Radoičić, R.: Rainbow solutions to the Sidon equation. Discrete Math. **308**(20), 4773–4778 (2008)
11. Graham, R.L., Rothschild, B.L., Spencer, J.H.: Ramsey Theory. Wiley, New York (1990)
12. Huicochea, M., Montejano, A.: The structure of rainbow-free colorings for linear equations on three variables in $\mathbb{Z}_p$. Integers Electron. J. Combin. Number Theory **15A**, A8 (2015)
13. Jungić, V.: Basics of Ramsey Theory. Routledge (2023)
14. Jungić, V., Licht (Fox), J., Mahdian, M., Nešetril, J., Radoičić, R.: Rainbow arithmetic progressions and anti-Ramsey results. Combin. Probab. Comput. **12**(5–6), 599–620 (2003)
15. Jungić, V., Nešetril, J., Radoičić, R.: Rainbow Ramsey theory. Integers Electron. J. Combin. Number Theory **5**(2), A9 (2005)
16. Landman, B.M., Robertson, A.: Ramsey Theory on the Integers, No. 24. American Mathematical Society, Providence (2004)
17. Llano, B., Montejano, A.: Rainbow-free colorings for $x + y = cz$ in Z_p. Discrete Math. **312**(17), 2566–2573 (2012)
18. Montejano, A., Serra, O.: Twenty Years of Progress in Arithmetic Rainbow Ramsey Theory (in preparation)
19. Montejano, A., Serra, O.: Rainbow-free 3-colorings of Abelian groups. Electron. J. Combin. **19**(1), P45 (2012)
20. Szemerédi, E.: On sets of integers containing no k elements in arithmetic progression. Acta Arithm. **27**, 199–245 (1975)

Dividing Lines Among Fields

Samaria Montenegro

Abstract In this article we will discuss some results on the classification of stable, simple, and NIP fields from the perspective of model theory. Additionally, we will explore recent work by Silvain Rideau-Kikuchi and the author, where a general class of fields (pseudo I-closed fields) was constructed providing new examples of NTP_2 fields. These fields satisfy a *local-global* principle for the existence of points on varieties with respect to models of certain theories. Finally, we will address open questions and outline future research directions.

Keywords Pseudo I-closed fields · NTP_2 theories · Classification of stable fields

1 Introduction

This is an article on model theory. Model theory is a branch of mathematical logic that explores the connection between formal theories (sets of statements in a formal language describing particular properties of mathematical structures) and their models (the structures in which the statements of the theory are true). In particular, model theory interacts strongly with other areas of mathematics, including algebra, geometry, and number theory.

In the late 1970s, Shelah published his book *Classification Theory* [1] which became one of the most important areas of study in current model theory. As Shelah proposed, classification theory studies first-order mathematical theories by identifying *dividing lines*, which refer to combinatorial configurations (like linear orders or trees) that their models can encode. Models of theories that encode these configurations tend to be more complex from a model-theoretic perspective. In contrast, theories having no model encoding these configurations admit a tractable description leading to a clearer understanding of them.

S. Montenegro (✉)
Universidad de Costa Rica, Escuela de Matemática-CIMPA, San José, Costa Rica
e-mail: samaria.montenegroguzman@ucr.ac.cr

P. E. Harris et al. (eds.), *Women in Mathematics in Latin America*,
Research Perspectives Ghent Analysis and PDE Center 14,
https://doi.org/10.1007/978-3-031-97857-9_11

One of the most important dividing lines corresponds to *stable* and *unstable* theories. Shelah in [1], and many others later generated significant results and techniques for analyzing types and models in stable theories. Within algebraic structures like groups or fields, these model-theoretic characteristics correlate with the algebraic properties inherent in the structure. Stable theories are often referred to as the *paradise of model theory*, due to their especially desirable and manageable properties.

Unfortunately, most structures studied in mathematics are not stable. Over the past thirty years, it has become evident that many tools used in stability theory can be generalized to larger classes of theories, a line of research known as *neo-stability theory*. Two important dividing lines in this context are *simple theories* and *theories lacking the independence property (NIP theories)*, both of which include stable theories.

Two main research directions exist in the model theory of fields. The first is to classify a given field K (such as $\mathbb{C}, \mathbb{R}, \mathbb{Q}_p$) according to these dividing lines to determine where K falls. The second is knowing that a field K belongs to such and such dividing line, what do we know about the structure of K (i.e. is algebraically closed, separably close, henselian, etc.)?

In the first part of the paper, we will define some of the principal classes of theories (stable, simple, and NIP) and explore the current conjectures on classifying fields within these categories, as well as the advancements achieved to date.

In the second part, we will define the class of *theories without the tree property of the second kind* (NTP_2 theories), this class was defined also by Shelah and contains strictly the class of simple and NIP theories. Among the main known examples of NTP_2 fields are PRC and PpC bounded fields (see [2]). In the search for new examples of NTP_2 fields, in [3], Rideau-Kikuchi and the author propose a unified framework for studying these fields, the class of *pseudo-I-closed fields*, where I is a set of enriched theories of fields. These fields verify a *local-global principle* for the existence of points on varieties with respect to models of the theories in I. This approach also enables a good description of some fields equipped with multiple V-topologies, particularly pseudo-algebraically closed fields with a finite number of V-topologies. At the end of the article, we will briefly discuss open questions and future projects. I would like to thank Silvain Rideau-Kikuchi and Pablo Cubides for their many comments and suggestions for this article.

2 Stable, NIP, and Simple Fields

In this section, we will provide the basic definitions of stable, simple, and NIP theories. We will also discuss existing conjectures regarding the classification of fields and some of the currently known results. We will assume that the reader has a basic understanding of model theory, see [4] for more details.

2.1 Stable Fields

Stable fields play a crucial role in understanding the deep connections between algebraic structures and logic, often intersecting with topics like algebraic geometry, the theory of algebraic groups, and Galois theory. Intuitively, a theory is *stable* if an order cannot be defined within this theory, more precisely:

Definition 1 Let T be an $\mathcal{L}$-theory. We call T *stable* if no $\mathcal{L}$-formula has the order property, i.e. there are no $\mathcal{M} \models T$, an $\mathcal{L}$-formula $\varphi(x, y)$ and sequences $(a_i)_{i\in\mathbb{N}}$ and $(b_i)_{i\in\mathbb{N}}$ such that: $\mathcal{M} \models \varphi(a_i, b_j)$ if and only if $i \leq j$.

A field K is called *stable* if its complete theory $Th(K)$ in the language of rings $\mathcal{L}_r = \{+, \cdot, 0, 1\}$ is stable.

It is well known that algebraically closed fields and separably closed fields are stable. However, it is still not certain what all stable fields are. In fact, it is an open question whether $\mathbb{C}(x)$ is stable or not. In the 1970s, Shelah made the following conjecture about the classification of these fields, which has been important for years in the development of model theory.

Conjecture 1 Every infinite stable field is separably closed.

In the process of attempting to prove this conjecture, the following significant advancements have been achieved:

1. Infinite ω-stable fields are algebraically closed (Macintyre [5]).
2. Infinite superstable fields are algebraically closed (Cherlin and Shelah [6]).
3. Infinite stable bounded field is separably closed (Poizat [7]).
4. Infinite stable fields of finite dp rank are algebraically closed (Halevi and Palacin [8]).
5. Infinite large stable fields are separably closed (Johnson et al. [9]).

2.2 NIP Fields

One of the most important classes in classification theory is the class of NIP theories, to find an introduction to NIP theories, the lector can read [10]. Intuitively, NIP theories are theories in which it is not possible to find a formula that can define arbitrary subsets of arbitrarily large finite sets, more precisely:

Definition 2 Let T be an $\mathcal{L}$-theory. An $\mathcal{L}$-formula $\phi(x, y)$ has the *independence property* (IP) if there is a model $\mathcal{M} \models T$, sequences $(b_l)_{l\in\mathbb{N}}$ in $\mathcal{M}^{|y|}$ and $(a_A)_{A\in\mathcal{P}(\mathbb{N})}$ in $\mathcal{M}^{|x|}$ such that $\mathcal{M} \models \phi(a_A, b_l)$ if and only if $l \in A$.

We say that T is NIP if no formula has IP.

A field K is called NIP if its complete theory $Th(K)$ in the language of rings $\mathcal{L}_r = \{+, \cdot, 0, 1\}$ is NIP.

All the stables fields are NIP, but this class contains new important examples such as real closed fields, p-adically closed fields, and algebraically closed valued fields.

In attempting to classify all NIP fields, Shelah conjectured the following:

Conjecture 2 Every infinite NIP field is separably closed, real closed, or admits a non-trivial Henselian valuation.

It is easy to see that the conjecture for NIP fields implies the conjecture for stable fields. One of the most significant recent advances in model theory, particularly in the classification of NIP fields, was made by Johnson in [11], who in particular demonstrates the conjecture for the case of dp-minimal fields, which are a special case of NIP fields.

2.3 Simple Fields

Simple theories include stable theories and they are particularly important because they allow much of the forking apparatus from stability theory to be extended and developed in a more general context. The reader can see [12] for more details.

Simple theories are defined as those in which it is impossible to encode a tree, more specifically:

Definition 3 Denote by $\mathbb{N}^{<\mathbb{N}}$ the set of finite sequences of $\mathbb{N}$ and by $\mathbb{N}^{\mathbb{N}}$ the set of infinite sequences of $\mathbb{N}$. Let T be an $\mathcal{L}$-theory. An $\mathcal{L}$-formula $\phi(x, y)$ has the *tree property (TP)*, if there are tuples $(a_\eta)_{\eta \in \mathbb{N}^{<\mathbb{N}}}$ and some $k \geq 2$ such that for all $\sigma \in \mathbb{N}^{\mathbb{N}}$, $\{\phi(x, a_{\sigma|_n}) : n < \mathbb{N}\}$ is consistent; but for all $\eta \in \mathbb{N}^{<\mathbb{N}}$ $\{\phi(x, a_{\eta \frown n}) : n \in \mathbb{N}\}$ is k-inconsistent. We say that T is *simple* if no formula has TP.

A field K is called *simple* if its complete theory $Th(K)$ in the language of rings $\mathcal{L}_r = \{+, \cdot, 0, 1\}$ is simple.

In the case of simple fields, there is a very important example, the case of pseudo finite fields, or more in general the pseudo algebraically closed fields (PAC fields). PAC fields were introduced by Ax in [13] and is one of the main examples for the development of unstable theories.

Definition 4 A PAC field is a field K such that any geometrically integral variety over K admits a K-rational point.

This notion has proven to be very fruitful. It was eventually reinterpreted as a special case of a wide range of local-global principles for the existence of rational points.

Duret in [14] shows that a PAC field K which is not an algebraically closed field is not NIP. This in particular implies that these fields are outside the classes of fields studied previously.

Chatzidakis and Pillay showed in [15] that the bounded PAC fields (i.e. if for any integer n, the field has only finitely many extensions of degree n) are simple.

Additionally, Chatzidakis proved in [16] that the condition of boundedness is always necessary for PAC fields to be simple.

These fields became the only known examples of infinite fields within the class of simple fields, which led to the following conjecture:

Conjecture 3 Infinite simple fields are bounded PAC.

In [17], Pillay and Walsberg showed that if K is large and simple, then there is a bounded PAC field L such that the absolute Galois group of L is isomorphic (as a topological group) to the absolute Galois group of K.

3 NTP_2 Fields

The class of NTP_2 theories is a generalization of both simple theories and NIP theories. This class has been extensively studied from the perspective of model theory, and although it is more general than the previous classes, it still possesses pleasant properties from the standpoint of model theory. The reader can refer to [18] for a detailed introduction to these theories. The formal definition of this class of theories is:

Definition 5 Let T be an $\mathcal{L}$-theory. An $\mathcal{L}$-formula $\phi(x, y)$ has TP_2 if there is a model $\mathcal{M} \models T$ and $(a_{lj})_{l,j \in \mathbb{N}} \in \mathcal{M}^{|y|}$ and $k \in \mathbb{N}$ such that:

1. $\{\phi(x, a_{lj})_{j \in \mathbb{N}}\}$ is k-inconsistent for all $l \in \mathbb{N}$.
2. For all $f : \mathbb{N} \to \mathbb{N}$, $\{\phi(x, a_{lf(l)}) : l \in \mathbb{N}\}$ is consistent.

A theory is called NTP_2 if no formula has TP_2. A field K is NTP_2 if its complete theory $Th(K)$ in the language of rings $\mathcal{L}_r = \{+, \cdot, 0, 1\}$ is NTP_2.

For NTP_2 fields, one of the most important examples is a generalization of PAC fields, allowing for either orders or p-adic valuations.

Definition 6 Let K be a field of characteristic zero.

1. K is *pseudo real closed* (PRC) if for every geometrically integral variety over K, if V has a smooth point in $\overline{K}$ for every real closure $\overline{K}$ of K, then V has an K-rational point.
2. K is *pseudo p-adically closed* (PpC) if for every geometrically integral variety over K, if V has a smooth point in $\overline{K}$, for every p-adic closure $\overline{K}$ of K, then V has an K-rational point.

Note that this definition is a clear generalization of PAC fields, and that the only thing we are adding is an additional hypothesis related to a condition that the variety has points in all real or p-adic closures. This is done because we want to preserve, in some way, the orders or p-adic valuations that exist. In [2] the author showed that the bounded PRC and PpC are NTP_2, this was a conjecture originally made by Chernikov, Kaplan, and Simon.

This naturally raises the question, is it possible to generalize this notion of fields to obtain new fields that are NTP_2, where we could, for example, combine valuations with orders? To answer this question, Rideau-Kikuchi and the author in [3] defined a more general class of fields, the PIC fields, which generalize the concept of being pseudo closed, but concerning some theories T_i's.

Let I be a finite set of theories which eliminate quantifiers in (disjoint) relational expansions of the language of rings. We denote by T_i the element $i \in I$ and $\mathcal{L}_i$ its language.

We will suppose for each $i \in I$ that T_i admits a definable Henselian V-topology τ_i and, for every $M_i \models T_i$, any $\mathcal{L}_i(M_i)$-definable set X_i has non-empty τ_i-interior in (the M_i-points of) its Zariski closure.

Let $\mathcal{L}_I = \bigcup_i \mathcal{L}_i$ and $T_I = \bigcup_i T_{i,\forall}$, i.e. T_I is the union of the universal part of each theory. Let us fix some $\mathcal{L}_I$-field $K \models T_I$ and $M_i \models T_i$ such that $K \subseteq M_i$, for all $i \in I$.

Definition 7 We say that K is *pseudo I closed (PIC)* if for every smooth geometrically integral variety over K, and every $U_i \subset V(M_i)$ non empty τ_i-open, $\bigcap_i U_i(K)$ is non empty.

Observe that by quantifier elimination in T_i, the topology induced by τ_i on K, does not depend on $L_i \models T_i$.

Note that PIC fields contain the class of PRC and PpC fields with finitely many orders or p-adic valuations.

Example 1

1. For each $i \in I$, let T_i be the theory of real closed fields in the language of rings with the added predicate $<_i$. Then PIC fields are PRC fields with at most $|I|$ distinct orders.
2. For each $i \in I$, let T_i be the theory of p-adically closed fields in Macintyre's language. Then PIC are PpC with at most $|I|$ distinct p-adic valuations.

This definition includes many other examples of fields that are neither PRC nor PpC, in particular, we can have examples where we can combine orders and valuations. For example, Theorem 4.1 of [19] implies that if K is real closed (respectively p-adically closed or algebraically closed), and the $(v_i)_{0<i\leq n}$ are distinct valuations, then K is PIC. In that case, we need to take T_0 the theory of real closed fields in the language of orders fields (respectively p-adically closed fields or algebraically closed valued fields (ACVF), and if $i \neq 0$, T_i is ACVF.

In studying the model theory of PIC fields, we first verify whether the known results for PRC and PpC fields can be generalized to this context. To achieve this, as with these fields, we will assume the hypothesis that the fields are bounded. For PRC and PpC fields, this hypothesis implies that there are only finitely many orders or respectively p-adic valuations. The bounded hypothesis is fundamental because it means that, with enough constants symbols in the language, we can have some control of the algebraic extensions of the field.

One of the first steps is to understand the definable sets. Just as in the cases of PRC and PpC fields, the existing topologies play a crucial role. Theorem 6.12 and

Corollary 6.13 of [3] show that every definable set closely approximates a quantifier-free set; specifically, we can find a quantifier-free set in which the definable set is dense with respect to the topology generated by all the τ_i's. We can also generalize the amalgamation theorems for types of PAC, PRC, and PpC fields, obtaining Theorem 7.1. of [3].

We aim to study PIC fields within the framework of classification theory. The initial reasonable hypothesis was that if the T_i theories were NIP (as real closed fields or p-adically closed fields), then PIC fields would be NTP$_2$. However, we have obtained a somewhat stronger result:

Theorem 1 *Suppose that for all $i \in I$, T_i is NTP$_2$, then any perfect bounded $K \models$ PIC also is.*

Additionally, as a corollary, we found a more natural algebraic example whose classification was previously unknown, leading to the following result: Corollary 7.10 of [3].

Corollary 1 *Let $K \models$ PAC be perfect and bounded and v_i be n independent non-trivial valuations on K. Then K is NTP$_2$.*

Some of these results in [3] were written in a more general form, and in some cases, stronger results were obtained. In particular, it is possible to calculate the burden concerning the burden of each of the theories T_i's. Also, forking is characterized in relation to forking in the theories T_i's.

Although there is currently no classification conjecture for NTP$_2$ fields, it is worth noting that no examples beyond those previously mentioned are known. While this is not sufficient to classify NTP$_2$ fields, we believe that PIC fields encompass many of these fields, particularly those with finite burden.

4 Future Work

Many open questions can be explored for PIC fields. The imaginaries of PAC, PRC, and PpC bounded fields have been studied in detail previously (see [20, 21] and [22]), but this has not yet been done for PIC fields.

Definable groups have also been studied in the cases of PAC and PRC fields (see [23] and [24]), but not for PpC fields or more in general for PIC fields.

However, it is very likely that for both cases (imaginaries and definable groups), the same strategy already known for PRC and PpC fields can be applied.

Also, the case of unbounded fields is even more complex. We know that boundedness is necessary for PAC fields to be simple or, more generally, for PRC and PpC fields to be NTP$_2$. It is also known that some unbounded PAC fields, despite not being simple, possess some favorable properties from a model-theoretic perspective (see [25]). However, this analysis has not yet been conducted for PRC and PpC fields, and even less so for PIC fields. This is currently a work in progress being developed in collaboration with Silvain Rideau-Kikuchi.

References

1. Shelah, S.: Classification theory and the number of nonisomorphic models. In: Studies in Logic and the Foundations of Mathematics, vol. 92, 2nd edn. North-Holland, Amsterdam (1990)
2. Montenegro, S: Pseudo real closed field, pseudo p-adically closed fields and NTP2. Annals Pure Appl. Logic **168**(1), 191–232 (2017)
3. Montenegro, S., Rideau-Kikuchi, S.: Pseudo T-closed fields. Model Theory. **4**(1), 1–35 (2025)
4. Tent, K., Ziegler, M.: A Course in Model Theory. Lecture Notes in Logic, vol. 40. Cambridge University Press, Cambridge, UK (2012)
5. Macintyre, A: ω_1-categorical fields. Fundam. Math. **70**(3), 253–270 (1971)
6. Cherlin, G., Shelah, S.: Superstable fields and groups. Annals Math. Logic **18**(3), 227–270 (1980)
7. Poizat, B.: Groupes stables, avec types génériques réguliers. J. Symbol. Logic **48**(2), MR 704088, 339–355 (1983)
8. Halevi, Y., Palacin, D.: The dp-rank of abelian groups. J. Symbol. Logic **84**, 957–986 (2019)
9. Johnson, W., Tran, C.-M., Walsberg, E., Ye, J.: Etale open topology and the stable field conjecture. J. Eur. Math. Soc. **26**(10), 4033–4070 (2024)
10. Simon, P.: Lecture Notes on NIP Theories. Lecture Notes in Logic, vol. 44. Cambridge University Press, Cambridge (2015)
11. Johnson, W.: Dp-finite fields VI: the dp-finite Shelah conjecture. arXiv:2005.13989 (2020)
12. Wagner, F.: Simple Theories. Mathematics and Its Applications, vol. 503. Kluwer Academic Publishers, Dordrecht (2000)
13. Ax, J.: The elementary theory of finite fields. Annals Math. Ser. 2 **88**, 239–271 (1968)
14. Duret, J.: Les corps faiblement algébriquement clos non séparablement clos ont la propriété d'indépendance. In: Pacholski et al. (eds.) Model Theory of Algebra and Arithmetic. Springer Lecture Notes, vol. 834, pp. 135–157. American Mathematical Society, Providence (1980)
15. Chatzidakis, Z., Pillay, A.: Generic structures and simple theories. Annals Pure Appl. Logic **95**, 71–92 (1998)
16. Chatzidakis, Z: Simplicity and independence for pseudo-algebraically closed fields. In: Cooper, S.B., Truss, J.K. (ed.) Models and Computability, London Math. Soc. Lect. Notes Series, vol. 259, pp. 41–61. Cambridge University Press, Cambridge (1999)
17. Pillay, A., Walsberg, E.: Galois groups of large simple fields. J. Model Theory **2**(2), 357–380 (2023)
18. Chernikov, A.: Theories without the tree property of the second kind. Annals Pure Appl. Logic **165**, 695–723 (2014)
19. Johnson, W.: Forking and dividing in fields with several orderings and valuations. J. Math. Logic **43**, Paper No. 2150025 (2022)
20. Hrushovski, E.: Pseudo-finite fields and related structures. In: Model Theory and Applications, Quad. Mat., Aracne, Rome, vol. 11, pp. 151–212 (2002)
21. Montenegro, S.: Imaginaries in bounded pseudo real closed fields. Annals Pure Appl. Logic **168**(10), 1866–1877 (2017)
22. Montenegro, S., Rideau-Kikuchi, S.: Imaginaries, invariant types and pseudo p-adically closed fields. Trans. Amer. Math. Soc. **374**(2), 803–828 (2021)
23. Hrushovski, E., Pillay, A.: Groups definable in local fields and pseudo-finite fields. Israel J. Math. **85**(1–3), 203–262 (1994)
24. Montenegro, S., Onshuus, A., Simon, P.: Stabilizers, NTP2 groups with f-generics nd PRC fields. J. Inst. Math. Jussieu **19**(3), 821–853 (2020)
25. Chatzidakis, Z.: Properties of forking in ω-free pseudo-algebraically closed fields. J. Symbol. Logic **67**(3), 957–996 (2002)

Frames in Shift Invariant Spaces of Weighted Mixed Lebesgue Spaces $L_\mu^{p,q}$

Rocío Balderrama and Carolina Mosquera

Abstract In this paper we consider (p, q)-frames in shift-invariant spaces

$$V_\mu^{p,q}(\Phi) = \left\{ f \in L_\mu^{p,q} : f = \sum_{i=1}^{r} \sum_{j_1 \in \mathbb{Z}} \sum_{j_1 \in \mathbb{Z}^d} d_i(j_1, j_2)\phi_i(. - j_1, . - j_2), \right.$$

$$\left. (d_i(j_1, j_2)) \in \ell_\mu^{p,q}(\mathbb{Z} \times \mathbb{Z}^d) \right\}.$$

of weighted mixed Lebesgue spaces $L_\mu^{p,q}(\mathbb{R} \times \mathbb{R}^d)$, where $\Phi = \{\phi_1, \cdots, \phi_r\}$. We prove the equivalence of the frame property and the closedness for the weighted shift-invariant subspace $V_\mu^{p,q}(\Phi)$.

Keywords Shift-invariant spaces · Weighted shift invariant spaces · Mixed Lebesgue space

1 Introduction

Frames were introduced by Duffin and Schaeffer for nonharmonic Fourier series in [9]. The interest in frame theory is due to its use in wavelets theory, time frequency analysis, and sampling theory (see for example [1, 4–7, 11, 12, 14]). In these contexts, frames have some additional structure, such as generating a shift invariant space. The notion of Riesz basis has been generalized from $L^2(\mathbb{R}^d)$ to

R. Balderrama (✉) · C. Mosquera
Departamento de Matemática, Facultad de Ciencias Exactas y Naturales, Universidad de Buenos Aires, Ciudad Universitaria, Pabellón I (C1428EGA), Ciudad de Buenos Aires, Argentina e IMAS–CONICET, Buenos Aires, Argentina
e-mail: rbalde@dm.uba.ar; mosquera@dm.uba.ar

P. E. Harris et al. (eds.), *Women in Mathematics in Latin America*,
Research Perspectives Ghent Analysis and PDE Center 14,
https://doi.org/10.1007/978-3-031-97857-9_12

$L^p(\mathbb{R}^d)$ and studied in the context of shift invariant spaces by Jia et al. in [10]. While the concept of frame in Banach spaces was studied in the 1990s, its analysis in $L^p(\mathbb{R}^d)$, for p not equal to 2, was studied by Aldroubi et al. in [2]. In this work, the authors, given a set of generators of a shift invariant space in $L^p(\mathbb{R}^d)$, mainly give necessary and sufficient conditions for the set of generators to constitute a p-frame, and to generate a closed shift invariant subspace of $L^p(\mathbb{R}^d)$.

Weighted shift invariant spaces $V_\mu^p(\Phi)$, $p \in [1, +\infty]$, where μ is a weight, were introduced for the non uniform sampling as a direct generalization of the space $V^p(\Phi)$ [2, 17]. The determination of p and the signal smoothness are used for the optimal compression and coding signals and images (see [8]). In [13], the authors show that the main result of [2] also holds in the case of weighted shift invariant spaces.

Some signals are time-varying in practice, meaning that the signals live in time-space domains at the same time. Mixed Lebesgue space is a suitable tool for modeling and measuring time-space signals, due to the separate integrability for different variables. In fact, the function spaces with mixed norms have practical significance and have rapidly been developed and applied in many fields of mathematics. For instance, in PDEs as an example, functions defined by spacial and time variables may belong to certain mixed-norm spaces. It is known that $L^{p,q}(\mathbb{R} \times \mathbb{R}^d)$ inherits many excellent properties of the traditional $L^p(\mathbb{R}^d)$ space. However, the order of integration in the definition of $L^{p,q}(\mathbb{R} \times \mathbb{R}^d)$ is not commutative. Therefore, research on some issues in mixed-norm Lebesgue spaces also brings various challenges. Recently, Yingchun et al. in [18], consider some generalizations of the results of [2] for the case of (p, q)-frames in shift invariant spaces in $L^{p,q}(\mathbb{R} \times \mathbb{R}^d)$.

The aim of our work is to show that the main result of [3, 13, 18] holds in the case of shift invariant spaces in weighted mixed Lebesgue spaces $L_\mu^{p,q}(\mathbb{R} \times \mathbb{R}^d)$. Following these works we show results needing additional arguments depending on the weights and the high dimensionality.

2 Notation, Definitions, and Preliminaries

Denote by $L^1_{loc}(\mathbb{R} \times \mathbb{R}^d)$ the space of measurable functions integrable over compact subsets of $\mathbb{R} \times \mathbb{R}^d$. For a nonnegative function $\omega \in L^1_{loc}(\mathbb{R} \times \mathbb{R}^d)$ we say that it is submultiplicative if $\omega((x_1, x_2) + (y_1, y_2)) \le \omega(x_1, x_2)\omega(y_1, y_2)$, for all $(x_1, x_2), (y_1, y_2) \in \mathbb{R} \times \mathbb{R}^d$ and a function μ on $\mathbb{R} \times \mathbb{R}^d$ is ω-moderate if $\mu((x_1, x_2) + (y_1, y_2)) \le C\omega(x_1, x_2)\mu(y_1, y_2)$ for all $(x_1, x_2), (y_1, y_2) \in \mathbb{R} \times \mathbb{R}^d$. Functions μ and ω are called weights. We assume that ω is a submultiplicative weight, continuous and symmetric and μ is ω-moderate. The standard class of weights on $\mathbb{R} \times \mathbb{R}^d$ are of the polynomial type $\omega_{s_1,s_2}(x_1, x_2) = (1 + |x_1|)^{s_1}(1 + |x_2|)^{s_2}$, with $s_1, s_2 > 0$.

Let $p, q \in [1, +\infty]$. Weighted $L^{p,q}$ spaces with moderate weights are translation invariant spaces (see [2]). Also, we consider weighted sequence spaces $\ell_\mu^{p,q}(\mathbb{Z}\times\mathbb{Z}^d)$ with ω-moderate weight μ.

Let

$$\mathcal{L}_\omega^{p,q} = \left\{ f : ||f||_{\mathcal{L}_\omega^{p,q}} = \left\| \sum_{j_1\in\mathbb{Z}} \left\| \sum_{j_2\in\mathbb{Z}^d} f(x_1+j_1, x_2+j_2)\omega(x_1,x_2) \right\|_{L^q_{x_2}([0,1]^d)} \right\|_{L^p_{x_1}([0,1]^d)} < \infty \right\}$$

and

$$W_\omega^{p,q} = \left\{ f : ||f||_{W^{p,q}} = \left(\sum_{j_1\in\mathbb{Z}} \left(\sum_{j_2\in\mathbb{Z}^d} \sup |f(x_1+j_1, x_2+j_2)|^p \omega(j_1,j_2)^p \right)^{q/p} \right)^{1/q} < \infty \right\}.$$

It is easy to see that $W_\omega^{1,1} \subset W_\omega^{p,q} \subset \mathcal{L}_\omega^{\infty,\infty} \subset \mathcal{L}_\omega^{p,q} \subset L_\omega^{p,q}$, $W_\omega^{p,q} \subset W_\mu^{p,q} \subset L_\mu^{p,q}$ and $L_\omega^{p,q} \subset L_\mu^{p,q}$ for $1 \le p < q \le \infty$. For $p = q = \omega = 1$, $L^{1,1} = \mathcal{L}^{1,1}$. In addition, $\ell_\omega^{1,q} \subset \ell_\omega^{p,q} \subset \ell_\mu^{p,q}$, for $1 < p < q \le +\infty$.

From [2] and [13] we have

1. If $f \in L_\mu^{p,q}$, $g \in L_\omega^{1,q}$ and $p \in [1,\infty]$ then $||f * g||_{L_\mu^{p,q}} \le ||f||_{L_\mu^{p,q}} ||g||_{L_\omega^{1,q}}$.
2. If $f \in L_\mu^{p,q}$, $g \in W_\omega^{1,q}$ and $p \in [1,\infty]$ then $||f * g||_{W_\mu^{p,q}} \le ||f||_{L_\mu^{p,q}} ||g||_{W_\omega^{1,q}}$.
3. If $c \in \ell_\mu^{p,q}$ and $d \in \ell_\omega^{1,q}$, then holds the inequality $||c * d||_{\ell_\mu^{p,q}} \le ||c||_{\ell_\mu^{p,q}} ||d||_{\ell_\omega^{1,q}}$

Let $D_1 = (d_1(j_1, j_2))_{(j_1,j_2)\mathbb{Z}\times\mathbb{Z}^d} \in \ell^{p_1,q_1}$ and $D_2 = (d_2(j_1, j_2))_{(j_1,j_2)\mathbb{Z}\times\mathbb{Z}^d} \in \ell^{p_2,q_2}$ with $1/p_1 + 1/p_2 \ge 1$ and $1/q_1 + 1/q_2 \ge 1$, then their convolution $D_1 * D_2$ is the sequence defined as

$$(D_1 * D_2)(j_1, j_2) = \sum_{(k_1,k_2)\in\mathbb{Z}\times\mathbb{Z}^d} d_1(j_1 - k_1, j_2 - k_2) d_2(k_1, k_2).$$

We observe that $D_1 * D_2 \in \ell^{r_1,r_2}$ with $\frac{1}{r_1} = \frac{1}{p_1} + \frac{1}{p_2} - 1$ and $\frac{1}{r_2} = \frac{1}{q_1} + \frac{1}{q_2} - 1$. Denote by $\mathcal{WC}_\mu^{p,q}$, $1 \le p, q < \infty$, the space of all 2π-periodic functions with their sequences of Fourier coefficients in $\ell_\mu^{p,q}$.

We observe that for 2π-periodic functions K_1, K_2 with D_1, D_2 their Fourier coefficients respectively, if $D_1 * D_2 \in \ell_\mu^{p,q}$, then $D_1 * D_2$ is the sequence of Fourier coefficients of the product $K_1 K_2$.

Lemma 1 *Let* $1 \le p, q \le +\infty$. *If* $f \in L_\mu^{p,q}(\mathbb{R}\times\mathbb{R}^d)$ *and* $g \in W_\omega^{1,1}(\mathbb{R}\times\mathbb{R}^d)$, *then*

$$\left\| \left\{ \int_{\mathbb{R}} \int_{\mathbb{R}^d} f(x_1,x_2) g(x_1 - j_1, x_2 - j_2) \omega(x_1,x_2)\, dx_1 dx_2 \right\}_{(j_1,j_2)\in\mathbb{Z}\times\mathbb{Z}^d} \right\|_{\ell_\mu^{p,q}}$$
$$\le \|f\|_{L_\mu^{p,q}} \|g\|_{W_\omega^{1,1}}.$$

For $\Phi = (\phi_1, \ldots, \phi_r) \in (\mathcal{L}_\mu^{p,q})^r$, let

$$V_\mu^{p,q}(\Phi) = \left\{ f \in L_\mu^{p,q} : f = \sum_{i=1}^{r} \sum_{j_1 \in \mathbb{Z}} \sum_{j_1 \in \mathbb{Z}^d} d_i(j_1, j_2)\phi_i(. - j_1, . - j_2), \right.$$

$$\left. \text{with } (d_i(j_1, j_2)) \in \ell_\mu^{p,q}(\mathbb{Z} \times \mathbb{Z}^d) \right\}.$$

the shift invariant space generated by Φ. By the above Lemma, the space $V_\mu^{p,q}(\Phi)$ is well defined.

In this paper we mainly study (p, q)-frames of the form $\{\phi_i(\cdot - j_1, \cdot - j_2)\colon (j_1, j_2) \in \mathbb{Z} \times \mathbb{Z}^d, 1 \leq i \leq r\}$ in shift-invariant subspaces $V_\mu^{p,q}(\Phi)$ of $L_\mu^{p,q}(\mathbb{R} \times \mathbb{R}^d)$, which is a direct generalization of [3] and [13].

Definition The family $\{\phi_i(\cdot - j_1, \cdot - j_2)\colon (j_1, j_2) \in \mathbb{Z} \times \mathbb{Z}^d, 1 \leq i \leq r\}$ is called a (p, q)-frame of $V_\mu^{p,q}(\Phi)$ if $\phi_i \in L_\mu^{p',q'}(\mathbb{R} \times \mathbb{R}^d)$ for all $1 \leq i \leq r$, and there exist positive constants A and B such that for any $f \in L_\mu^{p,q}(\mathbb{R} \times \mathbb{R}^d)$,

$$A\|f\|_{L_\mu^{p,q}} \leq \sum_{i=1}^{r} \|\{\langle f, \phi_i(\cdot - j_1, \cdot - j_2)\rangle\}_{(j_1, j_2) \in \mathbb{Z} \times \mathbb{Z}^d}\|_{\ell_\mu^{p,q}} \leq B\|f\|_{L_\mu^{p,q}},$$

where p' and q' are the conjugate numbers of p and q, respectively.

For any sequence $D = \{d(j_1, j_2)\}_{(j_1, j_2) \in \mathbb{Z} \times \mathbb{Z}^d} \in \ell_\mu^{p,q}(\mathbb{Z} \times \mathbb{Z}^d)$ and $f \in \mathcal{L}_\mu^{p,q}(\mathbb{R} \times \mathbb{R}^d)$, define the semi-convolution

$$f *' D = \sum_{j_1 \in \mathbb{Z}} \sum_{j_2 \in \mathbb{Z}^d} d(j_1, j_2) f(\cdot - j_1, \cdot - j_2).$$

It is easy to verify that $f *'$ is a continuous map from $\ell_\mu^{p,q}(\mathbb{Z} \times \mathbb{Z}^d)$ to $L_\mu^{p,q}(\mathbb{R} \times \mathbb{R}^d)$.

The following lemmas are useful for the proofs of the main results, of which only an outline will be sketched. Furthermore, they are important for the characterization of the spaces and to ensure that the work is self-contained.

Lemma 2 *Let* $1 \leq p, q \leq +\infty$. *If* $D \in \ell_\mu^{p,q}(\mathbb{Z} \times \mathbb{Z}^d)$ *and* $f \in \mathcal{L}_\mu^{p,q}(\mathbb{R} \times \mathbb{R}^d)$, *then*

$$\|f *' D\|_{L_\mu^{p,q}} \leq \|D\|_{\ell_\mu^{p,q}} \|f\|_{\mathcal{L}_\mu^{p,q}}.$$

Lemma 3 *Let* $1 \leq p, q \leq +\infty$. *Let* $f \in \mathcal{L}_\omega^{p,q}$. *Then for any* $D \in \ell_\mu^{1,1}$,

$$\|f *' D\|_{\mathcal{L}_\mu^{p,q}} \leq \|D\|_{\ell_\mu^{1,1}} \|f\|_{\mathcal{L}_\omega^{p,q}}.$$

Lemma 4 *Let $D \in \ell_\mu^{1,1}$ and $f \in \mathcal{L}_\mu^{p,q}$ $(1 \leq p, q \leq +\infty)$. Then*

$$\|f *' D\|_{\ell_\mu^{p,q}} \leq \|D\|_{\ell_\mu^{1,1}} \|f\|_{\mathcal{L}_\mu^{p,q}}.$$

Lemma 5 *Let $D \in \ell_\mu^{1,1}$ and $f \in W_\mu^{1,1}$. Then*

$$\|f *' D\|_{W_\mu^{1,1}} \leq \|D\|_{\ell_\mu^{1,1}} \|f\|_{W_\mu^{1,1}}.$$

Let be the function $\phi \in L^{2,2}(\mathbb{R} \times \mathbb{R}^d)$. Its Fourier transform is defined as

$$\widehat{\phi}(\xi, \tilde{\xi}) = \int_{\mathbb{R}} \int_{\mathbb{R}^d} \phi(x_1, x_2) e^{-i(x_1,x_2)\dot{(}\xi,\tilde{\xi})} \, dx_1 dx_2.$$

Let $\Phi = (\phi_1, \ldots, \phi_r)^t$ and $\Psi = (\psi_1, \ldots, \psi_s)^t$. If $\hat{\phi}_i(\xi, \tilde{\xi})\overline{\hat{\psi}_{i'}(\xi, \tilde{\xi})}$ is integrable for any $1 \leq i \leq r$ and $1 \leq i' \leq s$, we define the $r \times s$ matrix

$$\left[\hat{\Phi}, \hat{\Psi}\right](\xi, \tilde{\xi})$$
$$= \left(\sum_{j_1 \in \mathbb{Z}} \sum_{j_2 \in \mathbb{Z}^d} \hat{\phi}_i(\xi + 2j_1\pi, \hat{\xi} + 2j_2\pi) \overline{\hat{\psi}_{i'}(\xi + 2j_1\pi, \hat{\xi} + 2j_2\pi)} \right)_{1 \leq i \leq r, 1 \leq i' \leq s}.$$

Lemma 6 *Suppose that $\Phi, \Psi \in (\mathcal{L}_w^{2,2})^r$, then $[\widehat{\Phi}, \widehat{\Psi}](\xi, \tilde{\xi})$ is continuous.*

3 Main Results

We begin showing a generalization of a result of [15] about a equivalence of the ℓ_p-stability of the synthesis operator S_Φ for different $1 \leq p \leq +\infty$, and a corollary about well localization of the inverse of the synthesis operator S_Φ. First we need the following definitions.

For a continuous function f on $\mathbb{R}$, we define the modulus of continuity

$$\omega_\delta(f)(x) = \sup_{|y| \leq \delta} |f(x + y) - f(x)|,$$

a delicate tool in mathematical analysis to measure the regularity of a function.

We say that a discrete subset Λ of $\mathbb{R}^d$ is relatively-separated if

$$R(\Lambda) = \sup_{x \in \mathbb{R}^d} \sum_{\lambda \in \Lambda} \chi_{\lambda + [0,1)^d}(x) < \infty.$$

Theorem 1 *Let Λ be a relatively-separated subset of $\mathbb{R}^{d+1}$, $\Phi = \{\phi_\lambda : \lambda \in \Lambda\}$ be a family of functions with the property that*

$$\| \sup_{\lambda \in \Lambda} |\phi_\lambda(\cdot + \lambda)| \|_{W^{1,1}_\omega} < +\infty$$

and

$$\lim_{\delta \to 0} \| \sup_{\lambda \in \Lambda} |\omega_\delta(\phi_\lambda)(\cdot + \lambda)| \|_{W^{1,1}_\omega} = 0.$$

If the synthesis operator S_Φ has $\ell^{p,q}$-stability for some $1 \leq p, q \leq \infty$, then it has $\ell^{\tilde{p},\tilde{q}}$-stability for any $1 \leq \tilde{p}, \tilde{q} \leq \infty$.

We are interested in the particular case when $\Phi = \{\phi_1(\cdot - j), \cdots, \phi_r(\cdot - j)\}_{j \in \mathbb{Z}^{d+1}}$. We do not include the second hypothesis because $\lim_{\delta \to 0} \|\omega_\delta(f)\|_{W^{1,1}_\omega} = 0$ for any continuous function f in the Wiener amalgam space $W^{1,1}_\omega$.

Corollary 1 *Let $\Phi = (\phi_1, \cdots, \phi_r) \in (W^{1,1}_\omega)^r$, $1 \leq p, q \leq +\infty$ and μ is ω-moderate. Define the synthesis operator $S_\Phi : (\ell^{p,q}_\mu(\mathbb{Z} \times \mathbb{Z}^d))^r \to V^{p,q}_\mu(\Phi)$ by*

$$S_\Phi : c = \{c^i_{j_1,j_2}\}_{(j_1,j_2) \in \mathbb{Z} \times \mathbb{Z}^d, 1 \leq i \leq r} \mapsto \sum_{i=1}^{r} \sum_{(j_1,j_2) \in \mathbb{Z} \times \mathbb{Z}^d} c^j_i \phi_i(. - j_1, . - j_2).$$

If the synthesis operator S_Φ has $\ell^{p,q}_\mu$-stability for some $p, q \in [1, +\infty]$, i.e., there exists a positive constant C such that

$$C^{-1} \|c\|_{(\ell^{p,q}_\mu(\mathbb{Z} \times \mathbb{Z}^d))^r} \leq \|S_\Phi c\|_{L^{p,q}_\mu(\mathbb{R} \times \mathbb{R}^d)} \leq C \|c\|_{(\ell^{p,q}_\mu(\mathbb{Z} \times \mathbb{Z}^d))^r}, \tag{1}$$

for all $c \in (\ell^{p,q}_\mu(\mathbb{Z} \times \mathbb{Z}^d))^r$, then the synthesis S_Φ has $\ell^{\tilde{p},\tilde{q}}$-stability for any $\tilde{p}, \tilde{q} \in [1, +\infty]$.

It is easy to see that S_Φ has $\ell^{2,2}$-stability if and only if the matrix $A = (a(\lambda, \lambda'))_{\lambda, \lambda' \in \Lambda}$ has $\ell^{2,2}$-stability where $a(\lambda, \lambda') = \int_{\mathbb{R} \times \mathbb{R}^d} \phi_\lambda(x) \phi_{\lambda'}(x)\, dx$, for $\lambda, \lambda' \in \Lambda$. This observation together with the equivalence in the above theorem, and the Wiener's lemma in [16] for the Sjöstrand class of matrices leads the following proposition.

Proposition 1 *Let $\Phi = (\phi_1, \cdots, \phi_r) \in (W^{1,1}_\omega)^r$, $1 \leq p, q \leq +\infty$ and μ is ω-moderate. If the synthesis operator S_Φ has $\ell^{p,q}_\mu$-stability, then there exists another family $\Psi = (\psi_1, \cdots, \psi_r) \in (W^{1,1}_\omega)^r$ such that the inverse of the synthesis operator S_Φ is given by*

$$(S_\Phi)^{-1}(f) = \left\{ \int_{\mathbb{R}} \int_{\mathbb{R}^d} f(x_1, x_2) \psi_i(x_1 - j_1, x_2 - j_2)\, dx_1 dx_2 \right\}_{1 \leq i \leq r, (j_1, j_2) \in \mathbb{Z} \times \mathbb{Z}^d},$$

for all $f \in V^{p,q}_\mu(\Phi)$.

We are now ready to present our main theorem

Theorem 2 *Let* $\Phi = (\phi_1, \cdots, \phi_r) \in (W^{1,1}_\omega)^r$, $1 \le p, q \le +\infty$ *and* μ *is* ω-*moderate. Then the following statements are equivalent*

(i) $V^{p,q}_\mu(\Phi)$ *is closed in* $L^{p,q}_\mu$.
(ii) $\{\phi_i(\cdot - j_1, \cdot - j_2) : (j_1, j_2) \in \mathbb{Z}\times\mathbb{Z}^d,\ 1 \le i \le r\}$ *is a* (p, q)*-frame for* $V^{p,q}_\mu(\Phi)$.
(iii) *There exist a positive constant* $C > 0$ *such that*

$$C^{-1}[\widehat{\Phi}, \widehat{\Phi}](\xi, \tilde{\xi}) \le [\widehat{\Phi}, \widehat{\Phi}](\xi, \tilde{\xi})\overline{[\widehat{\Phi}, \widehat{\Phi}](\xi, \tilde{\xi})^T} \le C[\widehat{\Phi}, \widehat{\Phi}](\xi, \tilde{\xi}),$$
$$(\xi, \tilde{\xi}) \in [-\pi, \pi] \times [-\pi, \pi]^d.$$

(iv) *There exists a positive constant* B *(depending on* p, q *and* Φ*) such that*

$$B^{-1}||f||_{L^{p,q}_\mu} \le \inf_{f=\sum_{i=1}^r \phi_i *' D_i} \sum_{i=1}^r ||D_i||_{l^{p,q}_\mu} \le B||f||_{L^{p,q}_\mu}, \quad \text{for } f \in V^{p,q}_\mu(\Phi).$$

(v) *There exists* $\Psi = (\psi_1, \cdots, \psi_r) \in (W^{1,1}_\omega)^r$ *such that*

$$f = \sum_{i=1}^r \sum_{j_1 \in]Z} \sum_{j_2 \mathbb{Z}^d} \langle f, \psi_i(\cdot - j_1, \cdot - j_2)\phi_i(\cdot - j_1, \cdot - j_2)$$
$$= \sum_{i=1}^r \sum_{j_1 \in]Z} \sum_{j_2 \mathbb{Z}^d} \langle f, \phi_i(\cdot - j_1, \cdot - j_2)\psi_i(\cdot - j_1, \cdot - j_2), \quad f \in V^{p,q}_\mu(\Phi).$$

The following lemmas will be useful to prove Theorem 2. The proofs are similar to those of the corresponding lemmas from [18] hence, we will not include their proofs.

Lemma 7 *Let* $\Phi = (\phi_1, \cdots, \phi_r)^t \in \mathcal{L}^{2,2}_w$. *Then the following statements are equivalent to each other:*

1. *rank* $(\widehat{\Phi}(\xi + 2j_1\pi, \tilde{\xi} + 2j_2\pi))_{(j_1, j_2) \in \mathbb{Z}\times\mathbb{Z}^d}$ *is a constant function on* $\mathbb{R} \times \mathbb{R}^d$.
2. *rank* $[\widehat{\Phi}, \widehat{\Phi}](\xi, \tilde{\xi})$ *is a constant function on* $\mathbb{R} \times \mathbb{R}^d$.
3. *There exists a positive constant* C *independent of* ξ *and* $\tilde{\xi}$ *such that*

$$C^{-1}[\widehat{\Phi}, \widehat{\Phi}](\xi, \tilde{\xi}) \le [\widehat{\Phi}, \widehat{\Phi}](\xi, \tilde{\xi})\overline{[\widehat{\Phi}, \widehat{\Phi}](\xi, \tilde{\xi})^t} \le C[\widehat{\Phi}, \widehat{\Phi}](\xi, \tilde{\xi}),$$

for all $(\xi, \tilde{\xi}) \in [-\pi, \pi] \times [-\pi, \pi]^d$.

The following result provides a localization technique in Fourier domain. It allows us to replace locally the generator of size r by $\widehat{\Psi}_{1,(\lambda_1,\lambda_2)}$ of size k_0.

Lemma 8 *Let $\Phi \in \mathcal{L}_w^{2,2}$ satisfying rank $(\widehat{\Phi}(\xi + 2k_1\pi, \tilde{\xi} + 2k_2\pi))_{(k_1,k_2)\in\mathbb{Z}\times\mathbb{Z}^d} = k_0 \geq 1$ for all $(\xi, \tilde{\xi}) \in \mathbb{R} \times \mathbb{R}^d$. Then there exist a finite index set Λ, points $(\eta_{\lambda_1}, \eta_{\lambda_2}) \in [-\pi, \pi] \times [-\pi, \pi]^d$, a non singular 2π-periodic $r \times r$ matrix $P_{(\lambda_1,\lambda_2)}(\xi, \tilde{\xi})$ with all entries in the class $\mathcal{WC}_w^{1,1}$ and $(K_{\lambda_1}, K_{\lambda_2}) \subset \mathbb{Z} \times \mathbb{Z}^d$ with cardinality k_0 for all $(\lambda_1, \lambda_2) \in \Lambda$ and $\delta_{(\lambda_1,\lambda_2)} > 0$ small enough such that:*

(i) $\bigcup_{(\lambda_1,\lambda_2)\in\Lambda} B(\eta_{(\lambda_1,\lambda_2)}, \delta_{(\lambda_1,\lambda_2)}/2) \supset [-\pi, \pi] \times [-\pi, \pi]^d$, *where $B(x_0, \delta)$ denotes the open ball in $\mathbb{R} \times \mathbb{R}^d$ with center x_0 and radius $\delta > 0$.*

(ii)

$$P_{(\lambda_1,\lambda_2)}(\xi, \tilde{\xi})\widehat{\Phi}(\xi, \tilde{\xi}) = \begin{pmatrix} \widehat{\Psi}_{1,(\lambda_1,\lambda_2)}(\xi, \tilde{\xi}) \\ \widehat{\Psi}_{2,(\lambda_1,\lambda_2)}(\xi, \tilde{\xi}) \end{pmatrix}, \ \forall(\xi, \tilde{\xi}) \in \mathbb{R} \times \mathbb{R}^d$$

where $\Psi_{1,(\lambda_1,\lambda_2)}$ and $\Psi_{2,(\lambda_1,\lambda_2)}$ are functions form $\mathbb{R}^d$ to $\mathbb{C}^{k_0}$ and $\mathbb{C}^{r-k_0}$, respectively satisfying

$$\begin{aligned} &\text{rank } \widehat{\Psi}_{1,(\lambda_1,\lambda_2)}(\xi + 2k\pi)_{(k_1,k_2)\in(K_{\lambda_1},K_{\lambda_2})} \\ &\quad = k_0, \ \forall(\xi, \tilde{\xi}) \in B((\eta_{\lambda_1}, \eta_{\lambda_2}), 2\delta_{(\lambda_1,\lambda_2)}) \end{aligned}$$

and

$$\widehat{\Psi}_{2,(\lambda_1,\lambda_2)}(\xi, \tilde{\xi}) = 0, \ \forall(\xi, \tilde{\xi}) \in B(\eta_{(\lambda_1,\lambda_2)}, 2\delta_{(\lambda_1,\lambda_2)}) + 2\pi(\mathbb{Z} \times \mathbb{Z}^d).$$

Furthermore, there exist 2π-periodic C^∞ functions $h_{(\lambda_1,\lambda_2)}$, $(\lambda_1, \lambda_2) \in \Lambda$, on $\mathbb{R} \times \mathbb{R}^d$ such that $\sum_{(\lambda_1,\lambda_2)\in\Lambda} h_{(\lambda_1,\lambda_2)} \forall(\xi, \tilde{\xi}) = 1$, $(\xi, \tilde{\xi}) \in \mathbb{R} \times \mathbb{R}^d$, and supp$(h_{(\lambda_1,\lambda_2)}) \subset B(\eta_{(\lambda_1,\lambda_2)}, \delta_{(\lambda_1,\lambda_2)}) + 2\pi(\mathbb{Z} \times \mathbb{Z}^d)$.

Lemma 9 *Let $\phi \in \mathcal{L}_{\omega_{s_1,s_2}}^{p,q}$ if $1 \leq p, q \leq \infty$ and $\phi \in \mathcal{W}_{\omega_{s_1,s_2}}^{1,1}$ if $p = \infty$ or $q = \infty$. Assume that $\sum_{j_1\in\mathbb{Z}} \sum_{j_2\in\mathbb{Z}^d} \phi(\cdot - j_1, \cdot - j_2) = 0$. Then for any function h on $\mathbb{R} \times \mathbb{R}^d$ which satisfies*

$$|h(x_1, x_2)| \leq C(1 + |x_1|)^{-s_1-2}(1 + |x_2|)^{-s_2-d-1}$$

and

$$\begin{aligned} |h(x_1, x_2) - h(y_1, y_2)| &\leq C|x_1 - y_1||x_2 - y_2|(1 + \min(|x_1|, |y_1|))^{-s_1-2} \\ &\quad \times (1 + \min(|x_2|, |y_2|))^{-s_2-d-1} \end{aligned}$$

we have

$$\lim_{n\to\infty} 2^{-n(d+1)}\|\sum_{j_1\in\mathbb{Z}} \sum_{j_2\in\mathbb{Z}^d} h(2^{-n}j_1, 2^{-n}j_2)\phi(\cdot - j_1, \cdot - j_2)\|_{\mathcal{L}_{\omega_s}^{p,q}} = 0.$$

Now we are in condition to give the proof of our theorem.

Sketch of the Proof of Theorem 2 Since (i) implies that the synthesis operator S_{Φ} has $\ell_{\mu}^{p,q}$-stability we obtain statement $iv)$. In addition, if statement $iv)$ is satisfied, from Corollary 1 and after some computations we obtain the $\ell_{\mu}^{p,q}$-stability of the synthesis operator. We prove that, this is equivalent to (i), thus we have $(i) \iff (iv)$. Moreover, the $\ell_{\mu}^{p,q}$-stability implies (ii), obtaining $(iv) \iff (ii)$. From Lemma 1 we obtain the equivalence $(iv) \iff (v)$. To prove $(iii) \implies (v)$ it is necessary Lemma 8. Finally $(ii) \implies (iii)$ follows from Lemmas 7, 8, and 9. □

Remark 2 We would like to emphasize that the results presented in this article have the potential to be extended to the space $L^{(p_1,\ldots,p_r)}(\mathbb{R} \times \cdots \times \mathbb{R})$. This extension can be achieved with only minor modifications to the proofs that we have provided, allowing for a broader application of our work in related mathematical contexts.

References

1. Aldroubi, A., Feichtinger, H.: Exact iterative reconstruction algorithm for multivariate irregularly sampled functions in spline-like space: the L^p-theory. Proc. Am. Math. Soc. **126**, 2677–2686 (1998)
2. Aldroubi, A., Gröchening, K.: Non-uniform sampling and reconstruction in shift-invariant spaces. SIAM Rev. **43**(4), 585–620 (2001)
3. Aldroubi, A., Sun, Q., Tang, W.: p-frames and shift-invariant subspaces of L^p. J. Fourier Anal. Appl. **7**, 1–21 (2001)
4. Benedetto, J., Frazier, M.: Wavelets: Mathematics and Applications. CRC Press, Boca Raton (1994)
5. Benedetto, J., Li, S.: Multiresolution analysis frames with applications. In: ICASSP'93, Minneapolis, April 26–30, III, pp. 304–307 (1993)
6. Daubechies, I.: The wavelet transform, time-frequency localization and signal analysis. IEEE Trans. Inform. Theory **22**, 592–593 (1990)
7. Daubechies, I.: Ten Lectures on Wavelets. SIAM, Philadelphia (1992)
8. DeVore, R.A., Jawerth, B., Lucier, B.J.: Image compression through wavelet transform codings. IEEE Trans. Inform. Theory **38**, 719–746 (1992)
9. Duffin, R.J., Schaeffer, A.C.: A class of nonharmonic Fourier series. Trans. Am. Math. Soc. **72**, 341–366 (1952)
10. Jia, R.-Q., Micchelli, C.A.: Using the refinement equation for the construction of pre-wavelets II: power of two. In: Laurent, P.J., Le Méhaute, A., Schumaker, L.L. (eds.) Curve and Surface, pp. 209–246. Academic Press, New York (1991)
11. Mallat, S.: A Wavelet Tour of Signal Processing. Academic Press, New York (1998)
12. Meyer, Y.: Ondelettes. Hermann, Paris, Editeurs des Sciences et des Arts (1990)
13. Pilipovic, S., Simić, S.: Frames for weighted shift-invariant spaces, Mediterr. J. Math. **7**, 897–912 (2012)
14. Ron, A., Shen, Z.: Frames and stable bases for shift-invariant subspaces of $L^2(\mathbb{R}^d)$. Canad. J. Math. **47**, 1051–1094 (1995)
15. Shin, C., Sun, Q.: Stability of localized operators. J. Funct. Anal. **256**(8), 2417–2439 (2009)
16. Sun, Q.: Wiener's lemma for infinite matrices. Trans. Am. Math. Soc. **359**(7), 3099–3123 (2007)
17. Xian, J., Li, S.: Sampling set conditions in weighted multiply generated shift-invariant spaces and their applications. Appl. Comput. Harmon. Anal. **23**(2), 171–180 (2007)
18. Yingchun, J., Li, J.: (p, q)-frames in shift-invariant subspaces of mixed Lebesgue spaces $L^{p,q}(\mathbb{R} \times \mathbb{R}^d)$. arXiv:2001.08519v3 (2020)

Cremona Transformations of $\mathbb{P}^3$ Stabilizing Quartic Surfaces

Daniela Paiva

Abstract Any Cremona transformation of $\mathbb{P}^3$ stabilizing a smooth quartic surface $S \subset \mathbb{P}^3$ induces an automorphism on S. The converse problem of determining which automorphisms of S are restrictions of Cremona transformations of the ambient space $\mathbb{P}^3$ is still open. In this survey paper, we provide the answer of this problem for quartic surfaces with Picard rank two.

Keywords Cremona transformations · Quartic surfaces · Picard rank · Sarkisov program

1 Introduction

The Cremona group $\mathrm{Bir}(\mathbb{P}^3)$ is the group of birational self-maps of $\mathbb{P}^3$, so called *Cremona transformations* of $\mathbb{P}^3$. It is a classical and much-studied object in algebraic geometry. Given a smooth quartic surface $S \subset \mathbb{P}^3$, we define $\mathrm{Bir}(\mathbb{P}^3, S) \subset \mathrm{Bir}(\mathbb{P}^3)$ as the subgroup of all Cremona transformations that stabilize S, meaning that $\varphi \in \mathrm{Bir}(\mathbb{P}^3)$ and $\varphi(S) = S$. Given $\varphi \in \mathrm{Bir}(\mathbb{P}^3, S)$, its restriction to S induces an automorphism since S is a K3 surface. Therefore, we have a homomorphism of groups $\Psi : \mathrm{Bir}(\mathbb{P}^3, S) \longrightarrow \mathrm{Aut}(S)$.

The following problem, posed by Gizatullin, asks if the map Ψ is surjective.

Problem 1 (Gizatullin) Let $S \subset \mathbb{P}^3$ be a smooth quartic surface and $g \in \mathrm{Aut}(S)$ be an automorphism. Is g induced by a Cremona transformation of $\mathbb{P}^3$?

If $S \subset \mathbb{P}^3$ is a smooth quartic surface with Picard rank $\rho(S) = 1$, the answer is trivial since $\mathrm{Aut}(S) = \{1\}$ (see [10, §15, Corollary 2.12]). In the case of higher Picard rank, Gizatullin's problem was first addressed by Oguiso who constructed the following two examples. The first one is a smooth quartic surface $S \subset \mathbb{P}^3$ with

D. Paiva (✉)
Instituto de Matematica Pura e Aplicada, Rio de Janeiro, Brazil
e-mail: da.paiva@impa.br

P. E. Harris et al. (eds.), *Women in Mathematics in Latin America*,
Research Perspectives Ghent Analysis and PDE Center 14,
https://doi.org/10.1007/978-3-031-97857-9_13

$\rho(S) = 2$, $\mathrm{Aut}(S) \cong \mathbb{Z}$ and $\mathrm{Bir}(\mathbb{P}^3, S) = \{1\}$ (see [15]). In particular, the image of the restriction map Ψ is trivial. The second one is a smooth quartic surface $S \subset \mathbb{P}^3$ with $\rho(S) = 3$, $\mathrm{Aut}(S) \cong \mathbb{Z}_2 * \mathbb{Z}_2 * \mathbb{Z}_2$ and surjective restriction map Ψ (see [14]). Based on these examples, a natural question posed by Oguiso in [14] is:

Problem 2 (Oguiso) Is every automorphism of finite order of any smooth quartic surface $S \subset \mathbb{P}^3$ induced by a Cremona transformation of $\mathbb{P}^3$?

Our survey paper presents recent work on Problems 1 and 2, in collaboration with Ana Quedo [16], and with Carolina Araujo and Sokratis Zikas [2]. We focus on smooth quartic surfaces $S \subset \mathbb{P}^3$ with Picard rank $\rho(S) = 2$ and determine the image of the map Ψ. In particular, we obtain a negative answer to Oguiso's question.

For $S \subset \mathbb{P}^3$ with $\rho(S) = 2$, we define an invariant $r(S) > 0$, called the *discriminant* of S, which plays an important role (see ($\star$)). It carries information about the geometry of the surface and the existence of certain curves (see Sect. 2).

Our main result is the following. It is a combination of [2, Theorem 2] and [16, Theorem 17 and Corollary 18].

Theorem 1 *Let $S \subset \mathbb{P}^3$ be a smooth quartic surface with $\rho(S) = 2$.*

(1) If $r(S) > 57$ or $r(S) = 52$, then $\mathrm{Bir}(\mathbb{P}^3, S) = \{1\}$.
(2) If $r(S) \leq 57$, $r(S) \neq 52$ and S is very general, then the restriction homomorphism $\Psi : \mathrm{Bir}(\mathbb{P}^3, S) \to \mathrm{Aut}(S)$ *is surjective.*

Following Oguiso's idea, we combine the *theory of K3 surfaces and their automorphisms* and the *Sarkisov program* from birational geometry to prove Theorem 1. On the one hand, smooth quartic surfaces of $\mathbb{P}^3$ are K3 surfaces. From the *theory of K3 surfaces and lattices*, automorphisms of S can be read off from isometries of the rank two lattice $\mathrm{Pic}(S)$. This is the first main tool we use to investigate $\mathrm{Aut}(S)$. On the other hand, we use the *Sarkisov program and its volume preserving version* to investigate $\mathrm{Bir}(\mathbb{P}^3, S)$. The Sarkisov program (see [5] and [9]) allows us to factorize any Cremona transformation of $\mathbb{P}^3$ into simpler birational maps, called *Sarkisov links*. The volume preserving version (see [6]) allows us to factorize maps in $\mathrm{Bir}(\mathbb{P}^3, S)$ into *volume preserving* Sarkisov links.

We then briefly explain the key points of our work. First, we prove that any non-trivial finite order automorphism of S is an involution uniquely related with an ample divisor class of self-intersection 2 (see Proposition 1). Consequently, no non-trivial automorphisms can be induced by regular maps of $\mathbb{P}^3$, i.e., $\mathrm{Aut}(\mathbb{P}^3, S) = \{1\}$. Second, by the volume preserving Sarkisov program, the existence of a Cremona transformation $\varphi \in \mathrm{Bir}(\mathbb{P}^3, S) \setminus \mathrm{Aut}(\mathbb{P}^3, S)$ forces the existence of curves on S with specific arithmetic genus p_a and degree d (see Proposition 3). This property leads to restrictions on the Picard lattice $\mathrm{Pic}(S)$, which in turn depend only on the discriminant $r(S)$. Thus, under the assumptions of the first part of Theorem 1 we get that $\mathrm{Bir}(\mathbb{P}^3, S) = \mathrm{Aut}(\mathbb{P}^3, S)$. Next, when S satisfies the hypothesis of the second part of Theorem 1, we determine the automorphism group $\mathrm{Aut}(S)$ and find generators for it, by studying the action on $\mathrm{Pic}(S)$. Once we obtain the generators of

Aut(S), we construct the Cremona maps that realize them by exploring the Sarkisov links derived from the blow-up of $\mathbb{P}^3$ along curves of those specific types (p_a, d).

The structure of this paper is the following. In Sect. 2, we study automorphisms of K3 surfaces, focusing particularly on quartics in $\mathbb{P}^3$. In Sect. 3, we present the Sarkisov program and the main results we use. Finally, in Sect. 4, we provide the proof of the main result, along with a counterexample for Oguiso's Problem 2. Throughout this paper we always work over $\mathbb{C}$. All the varieties are projective. By a curve, we mean an irreducible and reduced curve.

2 Automorphisms of K3 Surfaces

We dedicate this section to the study of automorphisms of K3 surfaces via lattices. We refer to [10] for more details.

A *K3 surface* S is a smooth projective surface with $h^1(S, \mathcal{O}_S) = 0$ and trivial canonical bundle $\Omega_S^2 \cong \mathcal{O}_S$. Thus, there exists a unique (up to scalar) nowhere vanishing 2-form ω_S. Any smooth quartic surface of $\mathbb{P}^3$ is a K3 surface.

Let S denote a K3 surface. The second cohomology group $H^2(S, \mathbb{Z})$ equipped with the cup product is a lattice, i.e, a finitely generated free abelian group together with a non-degenerated symmetric bilinear form. In addition, it has a Hodge decomposition

$$H^2(S, \mathbb{Z}) \otimes \mathbb{C} \cong H^2(S, \mathbb{C}) = H^{2,0}(S) \oplus H^{1,1}(S) \oplus H^{0,2}(S),$$

where $H^{2,0}(S) = H^0(S, \Omega_S^2) \cong \mathbb{C}\omega_S$ and $H^{1,1}(S)$ is orthogonal to $H^{2,0}(S) \oplus H^{0,2}(S)$, under the extension of the cup product to $H^2(S, \mathbb{C})$. An automorphism $g \in \text{Aut}(S)$ induces a Hodge isometry g^* on $H^2(S, \mathbb{Z})$ which preserves ampleness, i.e., $g^*_{\mathbb{C}}(\mathbb{C}\omega_S) = \mathbb{C}\omega_S$ and g^*A is ample for an ample divisor A. Thus, g^* preserves the lattice $H^{1,1}(S) \cap H^2(S, \mathbb{Z})$, which is isomorphic to the *Picard lattice* Pic(S) by the Lefschetz theorem on (1, 1)-classes, as well as its orthogonal complement $T(S) := \text{Pic}(S)^{\perp}$, called the *transcendental lattice*. The following result is a strong consequence of the close relationship between K3 surfaces and lattices.

Theorem 2 (Global Torelli Theorem) *There is a bijection between automorphisms of a K3 surface S and Hodge isometries of $H^2(S, \mathbb{Z})$ preserving ampleness.*

For an automorphism $g \in \text{Aut}(S)$, since $g^*(\mathbb{C}\omega_S) = \mathbb{C}\omega_S$, there is $\lambda_g \in \mathbb{C}^*$ such that $g^*\omega_S = \lambda_g \omega_S$. If g has finite order n, λ_g is an n-th root of unity of $\mathbb{C}$. We consider the subgroup $G \subset \text{Aut}(S)$ of automorphisms g with $\lambda_g = \pm 1$. This group G is a finite index subgroup of Aut(S). If $g \in G$, $\lambda_g = 1$ (resp., $\lambda_g = -1$) if and only if $g^* = Id$ (resp., $g^* = -Id$) on $T(S)$ and hence also on $H^2(S, \mathbb{Z}) \setminus (\text{Pic}(S) \oplus T(S))$. Thus, Theorem 2 allows us to determine the group G from isometries of Pic(S) that can be extended as $\pm Id$ to $T(S)$.

Remark 1 The Picard group $\text{Pic}(S)$ of a K3 surface S is an even lattice of signature $(1, \rho(S) - 1)$, where $\rho(S)$ is its Picard rank. By Morrison [12, Corollary 2.9], for any even lattice L with signature $(1, \rho)$ with $\rho < 10$, there exists the $(20 - \rho)$-dimensional family of L-polarized K3 surfaces S. A *very general* member S in this family satisfies that $\text{Pic}(S) \cong L$ and $G = \text{Aut}(S)$ (see for instance [13, Theorem 10.1.2 c)]).

Automorphisms of Quartics with Picard Rank Two Now we turn to the following setting. Let $S \subset \mathbb{P}^3$ be a smooth quartic surface with Picard rank $\rho(S) = 2$. We can express the Picard lattice as $\text{Pic}(S) = \mathbb{Z}H \oplus \mathbb{Z}W$, where H is a hyperplane section of S. Then, the intersection matrix in this basis has the following form

$$\begin{pmatrix} 4 & b \\ b & 2c \end{pmatrix} \in M_{2\times 2}(\mathbb{Z}). \tag{$\star$}$$

We denote by $r = r(S) := b^2 - 8c > 0$ the *discriminant* of S. It satisfies that $r \equiv 0, 1, 4 \pmod 8$. For a divisor class $D = nH + mW$ on S, we get that $4D^2 = d^2 - rm^2$, where $d := D \cdot H$. When D is a curve of arithmetic genus p_a, $D^2 = 2p_a - 2$. Thus, the existence of divisors D with $D^2 = k$ is equivalent to the existence of integer solutions of the generalized Pell equation $x^2 - ry^2 = 4k$. In particular, the existence of curves on S of arithmetic genus p_a and degree d depends only on r.

In [3], prime-order automorphisms of a K3 surface are classified from a lattice perspective. If the quartic surface S admits an automorphism g of prime order $p \neq 2$, $\text{Pic}(S)$ must correspond to a lattice listed in [3], which is not the case here. Furthermore, if $p = 2$, there exists an ample class A with $A^2 = 2$ such that, for any $D \in \text{Pic}(S)$ satisfying $g^*D = D$, we have $D = kA$ for some $k \in \mathbb{Z}$. Conversely, any ample class A with $A^2 = 2$ defines a double cover $S \to \mathbb{P}^2$, naturally inducing an involution on S. Therefore, we have established the following result.

Proposition 1 *Let $S \subset \mathbb{P}^3$ be a smooth quartic surface with $\rho(S) = 2$. If $g \in \text{Aut}(S)$ has finite order then g is an involution. Moreover, there is a bijection between involutions of S and ample divisor classes $A \in \text{Pic}(S)$ with $A^2 = 2$.*

Any involution of S belongs to G. In [8, Proposition 5 and Corollary 1], Galluzzi et al. proved that there are just four possibilities for G. Combining this and Proposition 1, we get

Proposition 2 *Let $S \subset \mathbb{P}^3$ be a smooth quartic surface with $\rho(S) = 2$. Then, the subgroup $G \subset \text{Aut}(S)$ is isomorphic to*

$$\left.\begin{array}{l} \{1\} \\ \mathbb{Z}_2 \\ \mathbb{Z} \\ \mathbb{Z}_2 * \mathbb{Z}_2 \end{array}\right\} \textit{ if and only if } \begin{cases} \exists D \text{ with } D^2 \in \{-2, 0\} \text{ and } \nexists A \text{ ample with } A^2 = 2, \\ \exists D \text{ with } D^2 \in \{-2, 0\} \text{ and } \exists A \text{ ample with } A^2 = 2, \\ \nexists D \text{ with } D^2 \in \{-2, 0\} \text{ and } \nexists A \text{ ample with } A^2 = 2, \\ \nexists D \text{ with } D^2 \in \{-2, 0\} \text{ and } \exists A \text{ ample with } A^2 = 2. \end{cases}$$

Here D and A are divisors. Notice that the subgroup G, and in particular the existence of involutions of S, depends only on r.

3 The Sarkisov Program

In this section, we present the main definitions and results from birational geometry used in Sect. 4. For the terminologies and more details, we refer to [5, 6, 9].

The Minimal Model Program (MMP) classifies algebraic varieties up to birational equivalence. The *Sarkisov program* factorizes birational maps between certain outputs of the MMP into simpler birational maps. These outputs and the simpler maps are called *Mori fiber spaces* (Mfs) and *Sarkisov links* respectively. The factorization theorem was proved by Corti [5] in dimension 3 and Hacon and McKernan [9] in higher dimensions.

Theorem 3 (Sarkisov Program) *Every birational map between Mfs is a composition of Sarkisov links.*

The projective space $\mathbb{P}^3$ is a Mfs. Hence, due to the Sarkisov program any Cremona map $\varphi \in \mathrm{Bir}(\mathbb{P}^3)$ can be written as $\varphi\colon \mathbb{P}^3 = Y_0 \overset{\psi_1}{\dashrightarrow} Y_1 \overset{\psi_2}{\dashrightarrow} \cdots \overset{\psi_n}{\dashrightarrow} Y_n = \mathbb{P}^3$, where Y_i are Mfs and ψ_i are the Sarkisov links. Each such link involves birational operations such as divisorial contractions and small modifications (flips, flops and antiflips).

To fully address Problem 1, we use the following version of the Sarkisov program.

Volume Preserving Sarkisov Program This special version is a generalization where the varieties and maps have an additional structure: *Mori fibered Calabi-Yau pairs* and *volume preserving Sarkisov links*. It fits perfectly with the study of Gizatullin's problem since the pair $(\mathbb{P}^3, S)$ is a Calabi-Yau pair when $S \subset \mathbb{P}^3$ is a smooth quartic, and a Cremona transformation $\varphi \in \mathrm{Bir}(\mathbb{P}^3)$ belongs to $\mathrm{Bir}(\mathbb{P}^3, S)$ if and only if it is volume preserving.

More precisely, a pair (Y, D) is called *Calabi-Yau pair* if Y is a terminal projective variety and D is an effective Weil divisor on Y such that $K_Y + D \sim 0$ and (Y, D) has log canonical singularities. For any Calabi-Yau pair (Y, D) with $\dim Y = n$, there exists a unique rational form $\omega \in H^0(Y, \Omega^n_Y)$, up to scaling, such that $D + div(\omega) = 0$. We refer to ω as the *volume form* of the pair (Y, D). Let (Y, D) and (Y', D') be Calabi-Yau pairs, and ω and ω' their respective volume forms. A *volume preserving birational map* $\varphi : (Y, D) \dashrightarrow (Y', D')$ is a birational map $\varphi : Y \dashrightarrow Y'$ such that $\varphi^*\omega' = \lambda\omega$, for some $\lambda \in \mathbb{C}^*$.

Now, we state the volume preserving Sarkisov program which was established by Corti and Kaloghiros [6].

Theorem 4 (Volume Preserving Sarkisov Program) *Every volume preserving birational map between Mori fibered Calabi-Yau pairs is a composition of volume preserving Sarkisov links.*

In order to investigate Gizatullin's question for a smooth quartic $S \subset \mathbb{P}^3$, we study the elements of $\mathrm{Bir}(\mathbb{P}^3, S)$ by exploring its volume preserving Sarkisov decompositions. The first birational operation appearing in the first volume preserving Sarkisov link in the decomposition of any non-regular $\varphi \in \mathrm{Bir}(\mathbb{P}^3, S)$ is always a (volume preserving) divisorial contraction $X \to \mathbb{P}^3$. In fact, $X \to \mathbb{P}^3$ is the blow-up of $\mathbb{P}^3$ along a curve $C \subset S$ by [1, Proposition 3.1] and [17, Proposition 1.2]. In [4], Blanc and Lamy gave a complete list of pairs (g, d) for which the blow-up X of $\mathbb{P}^3$ along a general curve C of genus g and degree d is weak-Fano (i.e., $-K_X$ is big and nef), and described Sarkisov links initiated by the blowup of those curves. Using these facts we prove the following.

Proposition 3 *Let $S \subset \mathbb{P}^3$ be a smooth quartic surface with $\rho(S) = 2$ and $\varphi \in \mathrm{Bir}(\mathbb{P}^3, S) \setminus \mathrm{Aut}(\mathbb{P}^3, S)$. Then, any volume preserving Sarkisov decomposition of φ is initiated by the blow-up $X \to \mathbb{P}^3$ of a curve $C \subset S$, where X is weak Fano and (p_a, d) belongs to the set*

$$\left\{\begin{array}{c}(0,1),(0,2),(0,3),(0,4),(0,5),(0,6),(0,7),(1,3),(1,4),(1,5),(1,6),(1,7),\\(2,5),(2,6),(2,7),(2,8),(3,6),(3,7),(3,8),(4,6),(4,7),(4,8),(5,7),(5,8),\\(6,8),(6,9),(7,8),(7,9),(8,9),(9,9),(10,9),(10,10),(11,10),(14,11)\end{array}\right\}, \qquad (\dagger)$$

Here p_a denotes the arithmetic genus and d the degree of C. Conversely, if $C \subset S$ is a general curve with (p_a, d) in $(\dagger)$, then the blowup X of $\mathbb{P}^3$ along C is weak Fano and initiates a Sarkisov link.

4 Proof of the Main Result

We are now ready to sketch the proof of the main result. In this section, $S \subset \mathbb{P}^3$ is always a smooth quartic surface with $\rho(S) = 2$, discriminant $r = r(S)$ and Picard lattice $\mathrm{Pic}(S) = \mathbb{Z}H \oplus \mathbb{Z}W$, where H is the hyperplane section.

Proof of First Part of Theorem 1 Assume that $r > 57$ or $r = 52$. We first prove that $\mathrm{Bir}(\mathbb{P}^3, S) = \mathrm{Aut}(\mathbb{P}^3, S)$. Let $C = nH + mW$ be a curve on S that is not a complete intersection of S with another hypersurface $T \subset \mathbb{P}^3$, that is $m \neq 0$. Denote by p_a and d the arithmetic genus and degree of C, respectively. Thus, we have that $C^2 = 2p_a - 2$ and $8(p_a - 1) = 4C^2 = d^2 - rm^2$. It follows that r must divide $d^2 - 8(p_a - 1)$. This property is not satisfied by any pair in $(\dagger)$, and so (p_a, d) is not a pair in $(\dagger)$. Therefore, $\mathrm{Bir}(\mathbb{P}^3, S) = \mathrm{Aut}(\mathbb{P}^3, S)$, by Proposition 3. Next, we prove that $\mathrm{Aut}(\mathbb{P}^3, S) = \{1\}$.

Let $\varphi \in \mathrm{Aut}(\mathbb{P}^3, S)$ and assume that $g := \varphi|_S \neq Id_S$. Thus g is an involution of S by [15, Theorem 3.2] and Proposition 1. Since it is induced by a regular map of $\mathbb{P}^3$, $g^*H = H$ and $H = kA$, for some ample divisor A with $A^2 = 2$ and $k \in \mathbb{Z}$.

This is not possible since H is non-divisible in $\mathrm{Pic}(S)$. Hence, $g = Id_S$ and so, $\varphi = Id_{\mathbb{P}^3}$. □

Now, we exhibit a family of counter-examples for Oguiso's Problem 2.

Example 2 Let $b > 8$ be an integer and $S = S_b \subset \mathbb{P}^3$ be a generic smooth quartic surface with Picard lattice $\mathrm{Pic}(S) = \mathbb{Z}H \oplus \mathbb{Z}A$ and intersection matrix given by

$$\begin{pmatrix} 4 & b \\ b & 2 \end{pmatrix}.$$

Since $r = r(S) = b^2 - 8 > 57$, $\mathrm{Bir}(\mathbb{P}^3, S) = \{1\}$. Moreover, either $\mathrm{Aut}(S) = \mathbb{Z}_2$ or $\mathrm{Aut}(S) = \mathbb{Z}_2 * \mathbb{Z}_2$. Therefore, the involutions that generate $\mathrm{Aut}(S)$ as a free product, and hence all involutions of S, are not the restriction of a Cremona transformation of $\mathbb{P}^3$. Let us briefly show the existence of these surfaces. Notice that any rank two lattice with bilinear form given by the matrix above is even with signature $(1, 1)$. By Remark 1, there exists a general K3 surface S with $\mathrm{Pic}(S)$ as above. The fact that $r > 57$ implies that H is an ample divisor giving an embedding $S \hookrightarrow \mathbb{P}^3$ as a smooth quartic (see [16, Proposition 10]). Moreover, A is an ample divisor with $A^2 = 2$, and so, either $\mathrm{Aut}(S) = \mathbb{Z}_2$ or $\mathrm{Aut}(S) = \mathbb{Z}_2 * \mathbb{Z}_2$ by Proposition 2.

Proof Outline for the Second Part of Theorem 2 Assume that $r \leq 57, r \neq 52$ and S is very general. Note that $r \in \{9, 12, 16, 24, 25, 33, 36, 44, 49, 57\} \cup \{17, 41\} \cup \{28, 56\} \cup \{20, 32, 40, 48\}$ since $r \equiv 0, 1, 4 \pmod 8$ and [2, Proposition 2.17]. By Proposition 2 and Remark 1, the automorphism group $\mathrm{Aut}(S)$ can be determined:

$$\mathrm{Aut}(S) \cong \begin{cases} \{1\}, & \text{if } r \in \{9, 12, 16, 24, 25, 33, 36, 44, 49, 57\}; \\ \mathbb{Z}_2, & \text{if } r \in \{17, 41\}; \\ \mathbb{Z}_2 * \mathbb{Z}_2, & \text{if } r \in \{28, 56\}; \\ \mathbb{Z}, & \text{if } r \in \{20, 32, 40, 48\}. \end{cases}$$

When $\mathrm{Aut}(S) = \{1\}$ the statement follows trivially. It remains to find the generators for $\mathrm{Aut}(S)$ for each $r \in \{17, 41\} \cup \{28, 56\} \cup \{20, 32, 40, 48\}$, and construct the maps in $\mathrm{Bir}(\mathbb{P}^3, S)$ realizing them. The construction is the same for each of these three sets of values of r, and we explain it for some cases.

Assume $r = 17$. We can find a smooth curve $C \subset S$ with $(g, d) = (14, 11)$ such that $\mathrm{Pic}(S) = \mathbb{Z}H \oplus \mathbb{Z}C$. The automorphism group $\mathrm{Aut}(S) \cong \mathbb{Z}_2 = \langle g \rangle$ is generated by an involution. Following Lee's method (see [11, Theorem 1.2]), we obtain that the action of g on $\mathrm{Pic}(S)$ with respect to the basis $\{H, C\}$ is

$$g^* = \begin{pmatrix} 19 & 72 \\ -5 & -19 \end{pmatrix}.$$

By Proposition 3, the blow-up $p : X \to \mathbb{P}^3$ of $\mathbb{P}^3$ along C initiates a Sarkisov link φ_C which is indeed an involution of $\mathrm{Bir}(\mathbb{P}^3, S)$. This Sarkisov link φ_C is well-described in [18, Proposition 3.1 and Remark 3]. The Picard group $\mathrm{Pic}(X)$ of X has rank 2 and it is generated by H' and E, where H' is the pullback of H and E is the exceptional divisor of p. By [18, Proposition 3.4], φ_C fits to a small birational involution $\tilde{\varphi}_C$ of X and the action of $\tilde{\varphi}_C$ on $\mathrm{Pic}(X)$ is given by $\tilde{\varphi}_C^* H = 19H - 5E$ and $\tilde{\varphi}_C^* E = 72H - 19E$. This implies that $\varphi_C|_S = g$.

When $r(S) = 28$, there are exactly two classes of curves in $\mathrm{Pic}(S)$ of type $(10, 10)$. Let C and C' be smooth curves in each class, respectively. The Cremona transformations φ_C and φ'_C, constructed in the same way as the case $r(S) = 17$ above, are the ones realizing the two involutions that generate $\mathrm{Aut}(S) = \mathbb{Z}_2 * \mathbb{Z}_2$.

When $r = 20$, the automorphism group $\mathrm{Aut}(S) \cong \mathbb{Z}$ is generated by an infinite order automorphism g. Moreover, there is a smooth curve $C \subset S$ of type $(11, 10)$ such that $\mathrm{Pic}(S) = \mathbb{Z}H \oplus \mathbb{Z}C$. The class $4H - C$ in $\mathrm{Pic}(S)$ corresponds to a curve C^+ of type $(3, 6)$ in (†). In the basis $\{H, C\}$, the action g on $\mathrm{Pic}(S)$ is given by

$$g^* = \begin{pmatrix} 29 & 40 \\ -8 & -11 \end{pmatrix}.$$

The desired Cremona transformation ψ realizing it is the composition of the Sarkisov link φ_C initiated by the blow-up $p : X \to \mathbb{P}^3$ of C, and the Sarkisov link $\varphi_{C'}$ initiated by the blow-up $p' : X' \to \mathbb{P}^3$ of the curve $C' := \varphi_C(C^+)$. These two links have been studied in [4] and [7, Case 76], and the structure of each one is essentially the same as before. They are not elements of $\mathrm{Bir}(\mathbb{P}^3, S)$, but belong to $\mathrm{Bir}(\mathbb{P}^3)$.

Acknowledgments I would like to thank all my collaborators for the invaluable contribution to this work. I would also like to thank the Algebraic Geometry Working Group at the BIRS "Latin American and Caribbean Workshop on Mathematics and Gender" (Oaxaca 2022), where Gizatullin's problem was the focus. In particular, I sincerely appreciate the participants for their insightful discussions during and after the conference. Furthermore, I thank CAPES for its financial support.

References

1. Araujo, C., Corti, A., Massarenti, A.: Birational geometry of Calabi-Yau pairs and 3-dimensional Cremona transformations. arXiv e-prints (2023)
2. Araujo, C., Paiva, D., Zikas, S.: On Gizatullin's problem for quartic surfaces of Picard rank 2. arXiv e-prints (2024)
3. Artebani, M., Sarti, A.: K3 surfaces with non-symplectic automorphisms of prime order. Math. Z. **268**(1–2), 507–533 (2011)
4. Blanc, J., Lamy, S.: Weak Fano threefolds obtained by blowing-up a space curve and construction of Sarkisov links. Proc. Lond. Math. Soc. (3) **105**(5), 1047–1075 (2012)
5. Corti, A.: Factoring birational maps of threefolds after Sarkisov. J. Algebraic Geom. **4**(2), 223–254 (1995)
6. Corti, A., Kaloghiros, A.: The Sarkisov program for Mori fibred Calabi-Yau pairs. Algebr. Geom. **3**(3), 370–384 (2016)

7. Cutrone, J., Marshburn, N.: Towards the classification of weak Fano threefolds with $\rho = 2$. Cent. Eur. J. Math. **11**(9), 1552–1576 (2013)
8. Galluzzi, F., Lombardo, G., Peters, C.: Automorphs of indefinite binary quadratic forms and K3-surfaces with Picard number 2. Rend. Sem. Mat. Univ. Politec. Torino. **68**(1), 57–77 (2010)
9. Hacon, C., Mckernan, J.: The Sarkisov program. J. Algebraic Geom. **22**(2), 389–405 (2013)
10. Huybrechts, D.: Lectures on K3 Surfaces. Cambridge Studies in Advanced Mathematics 158. Cambridge University Press, Cambridge (2016)
11. Lee, K.: Automorphisms of K3 surfaces with Picard number two. Bull. Korean Math. Soc. **60**(6), 1427–1437 (2023)
12. Morrison, D.: On K3 surfaces with large Picard number. Invent. Math. **75**, 105–121 (1984)
13. Nikulin, V.: Factor groups of groups of automorphisms of hyperbolic forms with respect to subgroups generated by 2-reflections. J. Soviet Math. **22**, 1401–1475 (1983)
14. Oguiso, K.: Smooth quartic K3 surfaces and Cremona transformations, II. arXiv e-prints (2012)
15. Oguiso, K.: Quartic K3 surfaces and Cremona transformations. In: Arithmetic and Geometry of K3 Surfaces and Calabi-Yau Threefolds, Fields Inst. Commun, vol. 67, pp. 455–460. Springer, New York (2013)
16. Paiva, D., Quedo, A.: Automorphisms of quartic surfaces and Cremona transformations. J. Pure Appl. Algebra **229**(1), 107850 (2025)
17. Tziolas, N.: Terminal 3-fold divisorial contractions of a surface to a curve. I. Compos. Math. **139**(3), 239–261 (2003)
18. Zikas, S.: Rigid birational involutions of P3 and cubic threefolds. J. Éc. Polytech. Math. **10**, 233–252 (2023)

The Density of Gabor Systems in Expansible Locally Compact Abelian Groups

Victoria Paternostro, Emily J. King, and Rocío Nores

Abstract We investigate the reproducing properties of Gabor systems within the context of expansible groups. These properties are established in terms of density conditions. The concept of density that we employ mirrors the well-known Beurling density defined in Euclidean spaces, which is made possible due to the expansive structure. Along the way, for groups with a compact open subgroup, we demonstrate that modulation spaces are continuously embedded in Wiener spaces. Utilizing this result, we derive the Bessel condition of Gabor systems. We also provide a straightforward proof of the density result for Gabor frames, utilizing a comparison theorem for coherent frames.

Keywords Gabor systems · Density conditions · Modulation spaces

1 Introduction

A Gabor system is a set of functions in $L^2(\mathbb{R}^d)$ that are obtained by translating a single window in time and frequency along a set $\Lambda \subseteq \mathbb{R}^{2d}$. Their structure makes them of particular importance in many applications such as wireless communication, analysis and description of speech or music signals, and more (see, e.g., [13, 15, 25]). Because of this, it is important to study their reproducing properties, that is, to understand which properties of the generating window and the set Λ of time-frequency shifts guarantee the Gabor system to be an orthonormal basis, a Riesz basis, a frame, or a Bessel sequence. When the set Λ does not have an

V. Paternostro (✉) · R. Nores
Departamento de Matemática, Universidad de Buenos Aires, Instituto de Matemática "Luis Santaló" (IMAS-CONICET-UBA), Buenos Aires, Argentina
e-mail: vpater@dm.uba.ar; rnores@dm.uba.ar

E. J. King
Mathematics Department, Colorado State University, Fort Collins, CO, USA
e-mail: emily.king@colostate.edu

P. E. Harris et al. (eds.), *Women in Mathematics in Latin America*,
Research Perspectives Ghent Analysis and PDE Center 14,
https://doi.org/10.1007/978-3-031-97857-9_14

algebraic structure, these reproducing properties are stated in terms of its Beurling density.

Consider a set $\Lambda \subset \mathbb{R}^d$. The upper and lower Beurling densities of Λ, denoted by $D^+(\Lambda)$ and $D^-(\Lambda)$, respectively, capture its asymptotic behavior within balls of varying radii. In mathematical terms, they are given by:

$$D^+(\Lambda) = \limsup_{r\to\infty} \sup_{x\in\mathbb{R}^d} \frac{\#(\Lambda \cap B_r(x))}{r^d}, \quad \text{and} \quad D^-(\Lambda) = \liminf_{r\to\infty} \inf_{x\in\mathbb{R}^d} \frac{\#(\Lambda \cap B_r(x))}{r^d},$$

where $B_r(x)$ denotes the ball centered in x with radius r and # indicates the cardinality of a set. A thorough survey on the results about the density of Gabor systems is [18].

Beurling density is a concept that is also related to conditions for sampling and interpolation. See, for example, Landau's result in [23]. In order to explore the validity of Landau's result in the context of locally compact abelian groups that are compactly generated, in [17] Gröchenig et al. provide a definition of Beurling density by means of a comparison with a canonical lattice of reference. We refer to [1] for related results on the existence of sampling and interpolation sets near to the critical density in LCA groups.

In this paper we focus on studying Gabor systems in terms of density within the framework of expansible locally compact abelian groups, that is, locally compact abelian (LCA) groups which have a compact open subgroup and an expansive automorphism. (See Definition 1 for more details). One example of such an expansible LCA group is the space of p-adic numbers $\mathbb{Q}_p$.

This type of LCA group does not possess any lattices, and thus, the notion of density given in [17] needs to be reformulated. However, the presence of the expansive automorphism allows us to generalize the concept of density in the spirit of Beurling density in $\mathbb{R}^d$. A similar approach was given in [24].

Given an expansible LCA group G, a window $\varphi \in L^2(G)$ and a set $\Lambda \subseteq G \times \widehat{G}$, where $\widehat{G}$ denotes the Pontryagin dual group of G, we explore whether or not the Gabor system generated by φ with time-frequency shifts along Λ, $S(\varphi, \Lambda)$, is a Bessel sequence or a frame of $L^2(G)$ depending on the density of Λ.

We first move to a more abstract setting where we deal with unitary and projective representations of a locally compact group having a compact open subgroup. We based our analysis in the notion of modulation spaces introduced in [9] (see also [10–12]). We prove that the generalized wavelet transform (voice transform) induced by the (unitary or projective) representation continuously maps the modulation space of order p into the Wiener space $W(C, \ell^p)$. We then apply these results to the case were the projective representation is the one given by time-frequency translations, resulting in Gabor systems.

Returning to Gabor systems in expansive groups, we demonstrate that if $S(\varphi, \Lambda)$ is a Bessel sequence in $L^2(G)$, then Λ must have finite density. Conversely, when Λ has finite density and φ exhibits some decay, $S(\varphi, \Lambda)$ is a Bessel sequence. This latter result leverages the continuous embedding of modulation spaces into Wiener spaces.

Our findings serve as analogues for expansive LCA groups to [6, Theorem 3.1] and [18, Theorem 12], which are set in the Euclidean context.

Finally, we derive a density result for Gabor systems, providing a straightforward proof based on the comparison theorem [17, Theorem 4].

The article is structured as follows: In Sect. 2, we recall certain definitions about LCA groups and expansive automorphisms as well as what we will need about frame theory. We investigate in Sect. 3 the concept of density. We establish that it is (essentially) independent of the chosen automorphism used to define the density and show an equivalent condition for a set to have finite upper density. Moving to Sect. 4, we use the theory of modulation spaces introduced by Feichtinger [9] and developed further by Feichtinger and Gröchenig in [11, 12]. When the underlying locally compact group has a compact open subgroup, we establish one of the pivotal results of this work, namely, the continuous inclusion of modulation spaces into Wiener spaces through the generalized wavelet transform. Finally, in Sect. 5 we prove that the classical properties of Gabor systems hold true in our context.

2 Preliminaries

In this section we fix the context where we will work in. We will also recall some aspects of frame theory that we will need.

2.1 *LCA Groups*

A *locally compact abelian group* (*LCA group*) G is an abelian group that is also a locally compact topological space such that both multiplication and inversion are homeomorphisms of the space. We will always assume that the topology is Hausdorff. See, e.g., [21] for a general reference.

Given an LCA group G written additively, we denote by $\widehat{G}$ its Pontryagin dual group. By m_G we denote a Haar measure associated to G (with the desired normalization defined below). Since the dual of the dual group is topologically isomorphic to the original group, for $\xi \in \widehat{G}$ and $x \in G$ we write $\langle x, \xi \rangle$ to indicate the character ξ applied to x (i.e. $\xi(x)$) or the character x applied to ξ. For a subgroup $H \subseteq G$, its annihilator is denoted by $H^{\perp}$ and is defined as $H^{\perp} = \{\xi \in \widehat{G} : \langle h, \xi \rangle = 1, \forall h \in H\}$. It is well known that if H is closed, $H^{\perp}$ is a closed subgroup of $\widehat{G}$.

For a closed subgroup H of G, the dual of the quotient group G/H, $\widehat{G/H}$ is algebraically and topologically isomorphic to $H^{\perp}$ and $\widehat{H}$ is algebraically and topologically isomorphic to $\widehat{G}/H^{\perp}$.

When $H \subseteq G$ is a compact open subgroup, then so is $H^{\perp} \subseteq \widehat{G}$. As a consequence, the quotients G/H and $\widehat{G}/H^{\perp}$ are discrete abelian groups.

For an LCA group G with a compact open subgroup H, we consider the following normalization of the Haar measures involved: we fix m_G such that $m_G(H) = 1$ and $m_{\widehat{G}}(H^\perp) = 1$. As explained in [21, Comment (31.1)], this choice guarantees that the Fourier transform between $L^2(G)$ and $L^2(\widehat{G})$ is an isometry. We take $m_H = m_G|_H$, $m_{\widehat{H}} = m_{\widehat{G}}|_{H^\perp}$ and $m_{G/H}$ and $m_{\widehat{G}/H^\perp}$ to be the counting measures. Then, the Fourier transforms between $L^2(H)$ and $L^2(\widehat{G}/H^\perp)$ and between $L^2(G/H)$ and $L^2(H^\perp)$ are isometries.

2.2 Expansive Automorphisms

Let G be an LCA group. The group of homeomorphic automorphisms of G into itself is denoted by Aut(G). For a given $A \in \mathrm{Aut}(G)$, the measure μ_A defined by $\mu_A(U) = m_G(AU)$, where U is a Borel set of G, is a non-zero Haar measure on G. Therefore, there is a unique positive number $|A|$, the so-called *modulus of* A, such that $\mu_A = |A|m_G$.

For $A \in \mathrm{Aut}(G)$, there is an adjoint $A^* \in \mathrm{Aut}(\widehat{G})$ defined as $\langle Ax, \gamma\rangle = \langle x, A^*\gamma\rangle$ for all $x \in G$ and $\gamma \in \widehat{G}$. It holds that $|A^*| = |A|$.

We next present the definition of expansive automorphisms as given in [2], where they were used to define a wavelet theory over local fields. See also [24].

Definition 1 ([2, Definition 2.5]) Let G be an LCA group and $H \subseteq G$ a compact open subgroup, and let $A \in \mathrm{Aut}(G)$. We say that A is *expansive with respect to H* if the following two conditions hold:

1. $H \subsetneq AH$;
2. $\bigcap_{n\leq 0} A^n H = \{0\}$.

When H is fixed or clear from the context, we will simply say that A is expansive.

There exist many groups G with expansive automorphism A. We now give several examples.

Example 1 ([2, Example 2.10]) Let p be a prime number. Define the p-adic valuation over $\mathbb{Q}$ as $|p^r x|_p = p^{-r}$ for all $r \in \mathbb{Z}$ and all $x \in \mathbb{Q}$ such that the numerator and denominator of x are both relatively prime to p. Then $\mathbb{Q}_p$ is the completion of $\mathbb{Q}$ with respect to the p-adic valuation. Each element of $\mathbb{Q}_p$ may be represented as a Laurent series

$$\mathbb{Q}_p = \left\{ \sum_{n\geq n_0} a_n p^n : n_0 \in \mathbb{Z} \text{ and } a_n \in \{0, 1, \ldots, p-1\} \right\},$$

where addition "carries" rather than is modular. That is,

$$(p-1)p + 1p = p^2 \neq 0p.$$

Then $\mathbb{Q}_p$ is a locally compact abelian group with group operation addition and topology defined via the p-adic valuation. The Laurent series are not formal as they converge in the topology.

Then the p-adic integers $\mathbb{Z}_p$, defined as the set of power series

$$\mathbb{Z}_p = \left\{ \sum_{n\geq 0} a_n p^n : a_n \in \{0, 1, \ldots, p-1\} \right\},$$

is a compact open subset of $\mathbb{Q}_p$. Further characterizations of $\mathbb{Z}_p$ are that $\mathbb{Z}_p$ is the unit ball (with respect to p-adic valuation) of $\mathbb{Q}_p$ and the closure of $\mathbb{Z}$ in $\mathbb{Q}_p$.

The p-adic numbers are self-dual. Let $\{\cdot\} : \mathbb{Q}_p \to \mathbb{Q}_p$ be defined as

$$\left\{ \sum_{n\geq n_0} a_n p^n \right\} = \sum_{n=n_0}^{\max\{0,n_0\}} a_n p^n.$$

Then each $y \in \mathbb{Q}_p$ defines an element of $\widehat{\mathbb{Q}}_p$ as $\langle \cdot, y\rangle = \exp(2\pi i\, \{\cdot y\})$.

If we consider $A : \mathbb{Q}_p \to \mathbb{Q}_p$ to be the morphism given by $Ax = p^{-1}x$, then A is an automorphism of $\mathbb{Q}_p$, and it is easy to see that is expansive with respect to $\mathbb{Z}_p$.

Example 2 ([2, Example 2.11]) Let p be a prime number, where $\mathbb{F}_p$ is the field of order p. The additive group of the field $\mathbb{F}_p((t))$ of formal Laurent series in variable t:

$$\mathbb{F}_p((t)) = \left\{ \sum_{n\geq n_0} a_n t^n : n_0 \in \mathbb{Z} \text{ and } a_n \in \mathbb{F}_p \right\},$$

where addition is modular rather than “carries”:

$$(p-1)t + 1t = 0t = 0$$

is an LCA group with respect to the topology defined from an analog of the p-adic valuation. The set of formal power series

$$\mathbb{F}_p[[t]] = \left\{ \sum_{n\geq 0} a_n t^n : a_n \in \mathbb{F}_p \right\}$$

is a compact open subgroup. One possible expansive automorphism is multiplying by t^{-1}. However, the structure of $\mathbb{F}_p((t))$ yields a richer collection of automorphisms than for $\mathbb{Q}_p$.

Both classes of the LCA groups above are also fields. Further examples may be formed by considering the additive groups of finite field extensions or vector spaces over the above examples.

Example 3 ([2, Example 2.14]) Let G_1 be an LCA group with a compact open subgroup H_1 and G_2 be a nontrivial discrete abelian group. If we consider $G = G_1 \times G_2$, then G is an LCA group with a compact open subgroup $H = H_1 \times \{0\}$. Further, the annihilator $H^\perp$ is $H_1 \times \widehat{G_2}$ and if A_1 is an automorphism of G_1 which is expansive with respect to H_1, then $A = A_1 \times e_{G_2}$, for e_{G_2} the identity in G_2, is an automorphism of G which is expansive with respect to H. However, in this case, the union of all positive iterates $A^n H$ can not cover G since $AH = A_1 H_1 \times \{0\}$.

Following [24], we shall call an LCA group which admits an expansive automorphism an *expansible* group. Expansiveness can also be characterized by the action of the adjoint automorphism as the next lemma shows, whose proof can be found in [2, Lemma 2.6].

Lemma 1 *Let G be an LCA group with a compact open subgroup $H \subseteq G$, and let $A \in \mathrm{Aut}(G)$.*

(i) $H \subseteq AH$ *if and only if* $H^\perp \subseteq A^* H^\perp$.
(ii) $H \subsetneq AH$ *if and only if* $H^\perp \subsetneq A^* H^\perp$.
(iii) If $H \subseteq AH$, *then*

$$\bigcap_{n \le 0} A^n H = \{0\} \iff \bigcup_{n \ge 0} A^{*n} H^\perp = \widehat{G}. \tag{1}$$

Given an LCA group G with a compact open subgroup H and expansive automorphism $A : G \to G$, we define for $n \in \mathbb{Z}$ and $x \in G$,

$$Q_n(x) = x + A^n H.$$

One may think of this as the "ball" with "center" x and "radius" $|A|^n$, keeping in mind that if $y \in Q_n(x)$ then $Q_n(x) = Q_n(y)$, so the choice of "center" is not unique. In the case that $G = \mathbb{Q}_p$, $H = \mathbb{Z}_p$, and A is multiplication by $1/p$, the $Q_n(x)$ are precisely the balls in the metric induced by the p-adic valuation. Note that each $Q_n(x)$ is a compact open subset of G and also a coset of $A^n H$ in G. Moreover, by [20, Theorem 4.5], $\{Q_n(x)\}_{n \in \mathbb{Z}, x \in G}$ is a basis for the topology of G.

In this paper, we will be dealing with the specific scenario of forming "balls" in $G \times \widehat{G}$, where G satisfies the hypothesis above. In that case, we write for $n \in \mathbb{Z}$ and $(x, \gamma) \in G \times \widehat{G}$

$$\begin{aligned} Q_n(x, \gamma) &= (x, \gamma) + \left(A^n \otimes (A^*)^n\right)(H \times H^\perp) \\ &= \left(x + A^n H\right) \times \left(\gamma + (A^*)^n H^\perp\right) = Q_n(x) \times Q_n(\gamma). \end{aligned} \tag{2}$$

Note that in this case, the "radius" of $Q_n(x, \gamma)$ is $(|A||A^*|)^n = |A|^{2n}$.

When A is expansive with respect to H and A^* is expansive with respect to $H^\perp$, we have that $\{Q_n(x, \gamma)\}_{n\in\mathbb{Z}, x\in G, \gamma\in\widehat{G}}$ is a basis for the topology of $G \times \widehat{G}$. This is because for $n \leq m$, $x \in G$ and $\gamma \in \widehat{G}$, $Q_n(x) \times Q_m(\gamma) \subseteq Q_m(x, \gamma)$.

2.3 Frames and Riesz Bases

Let $\{\varphi_i\}_{i\in I}$ be a family of elements in a separable Hilbert space $\mathcal{H}$. It is said that $\{\varphi_i\}_{i\in I}$ is a *frame for* $\mathcal{H}$ if there exist two positive constants $\alpha \leq \beta$ such that

$$\alpha\|f\|^2 \leq \sum_{i\in I} |\langle f, \varphi_i\rangle|^2 \leq \beta\|f\|^2 \quad \forall f \in \mathcal{H}. \tag{3}$$

The constants α, β are called *frame bounds*. The *frame operator* defined as $Sf = \sum_{i\in I}\langle f, \varphi_i\rangle\varphi_i$ for $f \in \mathcal{H}$ is a bounded, invertible, and positive operator from $\mathcal{H}$ onto itself. This provides the well known frame decomposition

$$f = S^{-1}Sf = \sum_{i\in I}\langle f, \varphi_i\rangle\phi_i \quad \forall f \in \mathcal{H},$$

where $\phi_i = S^{-1}\varphi_i$. The family $\{\phi_i\}_{i\in I}$ is also a frame for $\mathcal{H}$, which is called the *canonical dual frame*, and has frame bounds $\beta^{-1} \leq \alpha^{-1}$. Any other frame $\{\widetilde{\phi}_i\}_{i\in I}$ satisfying $f = \sum_{i\in I}\langle f, \varphi_i\rangle\widetilde{\phi}_i \quad \forall f \in \mathcal{H}$, is called a *dual frame*, and it is well known that a frame can have dual frames besides the canonical one (typically infinitely many). For a general reference on frame theory, see, e.g., [5] and the references therein.

Riesz bases are special cases of frames and can be characterized as those frames which are biorthogonal to their canonical dual frame, i.e., such that $\langle\varphi_i, \phi_j\rangle = \delta_{ij}$.

A family $\{\varphi_i\}_{i\in I}$ which satisfies the right inequality in (3) (but possibly not the left) is called a *Bessel sequence* for $\mathcal{H}$.

3 Density with Respect to Expansive Automorphisms

In this section, using the balls defined above, we will consider the concept of density, which will extend that of the well known Beurling density for Euclidean spaces $\mathbb{R}^d$. This concept has been considered before in LCA groups that are compactly generated in [1, 17], however, with another approach. In [24] the authors work with the same density as here, but they consider a slightly different notion of expansive automorphism. In fact, their automorphism only satisfies condition (2) of Definition 1.

Definition 2 Let G be an LCA group, $H \subseteq G$ a compact open subgroup and $A \in \text{Aut}(G)$ expansive with respect to H. For $n \in \mathbb{Z}$, a set (countable or uncountable) $\Lambda \subseteq G$ is said to be *(A, n)-uniformly separated* if $\#\{\Lambda \cap Q_n(x)\} \leq 1$ for all $x \in G$. We say that Λ is simply *uniformly separated* if it is (A, n)-uniformly separated for some $n \in \mathbb{Z}$. Additionally, Λ is said to be *A-separated* if it is a finite union of uniformly separated sets.

Recall that if G is an LCA group and $H \subseteq G$ is a subgroup, a *section* of a quotient group G/H is a measurable set of representatives, and it contains exactly one element of each coset. As stated above, if $H \subseteq G$ is a compact open subgroup, since $H^{\perp}$ is also compact, then G/H is a discrete group. Thus, every section $C \subseteq G$ for the quotient G/H must be discrete as well. This is because for $x \in C$, $x = C \cap (x + H)$ and then, since $x + H$ is an open set in G, C is discrete with respect to the topology of G.

With this in mind, we can say that Λ is (A, n)-uniformly separated if and only if $\#\{\Lambda \cap Q_n(A^n c)\} \leq 1$ for all $c \in C$. This is a direct consequence of the fact that for every $n \in \mathbb{Z}$, $\{Q_n(A^n c)\}_{c \in C}$ is a partition of G, that is, $G = \bigcup_{c \in C} Q_n(A^n c)$ where the union is disjoint, and that $Q_n(A^n c) = Q_n(x)$ for every $x \in Q_n(A^n c)$.

Definition 3 Let G be an LCA group, $H \subseteq G$ a compact open subgroup and $A \in \text{Aut}(G)$ expansive with respect to H. For a subset $\Lambda \subseteq G$, the *upper and lower Beurling densities of* Λ are defined by

$$D_A^+(\Lambda) := \limsup_{n \to +\infty} \frac{1}{|A|^n} \max_{x \in G} \#\{\Lambda \cap Q_n(x)\},$$

and

$$D_A^-(\Lambda) := \liminf_{n \to +\infty} \frac{1}{|A|^n} \min_{x \in G} \#\{\Lambda \cap Q_n(x)\},$$

respectively. If $D_A^+(\Lambda) = D_A^-(\Lambda)$ we say that Λ has *uniform density* $D_A(\Lambda) = D_A^+(\Lambda) = D_A^-(\Lambda)$.

The analogy with the Beurling density defined in $\mathbb{R}^d$ [6, 18, 24] is clear from the definition noting that, since m_G is invariant under translations and $m_G(H) = 1$, we have that $|A|^n = m_G(A^n H) = m_G(Q_n(x))$ for any $x \in G$. Notice that, in the definition, it is not assumed Λ to be discrete. However, most of the applications will require that Λ is a uniformly separated set and therefore, discrete.

At first glance, the Beurling density seems to depend on the automorphism. However, if both the automorphism and their adjoint are expansive, the density becomes independent of the automorphism choice, as we show next. First, we observe some properties about sections that will be useful.

Remark 1 Given $A \in \text{Aut}(G)$ expansive with respect to a compact open subgroup H of G, the quotient AH/H must be finite. This is because AH is compact and then, it may be covered by finite (disjoints) cosets $x + H$. Let $C_0 \subseteq AH$ be a finite section for AH/H. Therefore, $AH = \bigcup_{c_0 \in C_0} H + c_0$, and then we have that $m_G(AH) = \#C_0$ which implies $|A| = \#C_0$. Moreover, since $AH \subseteq G$ is also a compact open subgroup, we can take C_1 a discrete section for G/AH. An easy computation shows that the set $C := C_0 + C_1$ must be a (discrete) section for G/H. From now on, we will consider sections for G/H of this form.

Lemma 2 *Let G be an LCA group, $H \subseteq G$ a compact open subgroup and A, $B \in \mathrm{Aut}(G)$ be expansive with respect to H such that A^* and B^* are expansive with respect to $H^{\perp}$ as well. Then, for every subset $\Lambda \subseteq G$,*

$$D_A^+(\Lambda) = D_B^+(\Lambda) \quad \textit{and} \qquad D_A^-(\Lambda) = D_B^-(\Lambda).$$

Proof Because of the expansiveness of A^* we know by (1) that $G = \bigcup_{n \geq 0} A^n H$. Let $n \in \mathbb{N}$ be fixed. Then there exists $k_0 = k_0(n) \in \mathbb{N}$ such that $B^n H \subseteq A^{k_0} H$ because $B^n H$ is compact and $\{A^k H\}_{k \in \mathbb{N}}$ is an open cover of nested sets.

Let $\{x_i\}_{i \in I} \subseteq A^{k_0} H$ be a section of $A^{k_0} H / B^n H$. Then

$$A^{k_0} H = \bigcup_{i \in I} (B^n H + x_i),$$

and for $x \in G$,

$$Q_{k_0}^A(x) = A^{k_0} H + x = \bigcup_{i \in I} (B^n H + x_i) + x = \bigcup_{i \in I} Q_n^B(x_i + x).$$

By taking intersections with Λ and taking the maximum cardinality over G, we obtain the following inequality:

$$\begin{aligned} \max_{x \in G} \# \left\{ Q_{k_0}^A(x) \cap \Lambda \right\} &= \max_{x \in G} \# \left\{ \bigcup_{i \in I} Q_n^B(x_i + x) \cap \Lambda \right\} \\ &\leq \sum_{i \in I} \max_{x \in G} \# \left\{ Q_n^B(x_i + x) \cap \Lambda \right\} \\ &= \sum_{i \in I} \max_{x \in G} \# \left\{ Q_n^B(x) \cap \Lambda \right\} = (\#I) \max_{x \in G} \# \left\{ Q_n^B(x) \cap \Lambda \right\}. \end{aligned}$$

Using Remark 1 and the fact that A^m is still an expansive automorphism with respect to H for every $m \in \mathbb{N}$, we can easily deduce that $\# \frac{A^m H}{H} = |A^m| = |A|^m$. Furthermore, $\#(A^{k_0} H/H) = \#(A^{k_0} H / B^n H) \#(B^n H/H)$ and then

$$\#I = \#(A^{k_0} H / B^n H) = \frac{|A|^{k_0}}{|B|^n},$$

Finally, we obtain

$$\frac{\max_{x\in G}\#\{Q^A_{k_0}(x)\cap\Lambda\}}{|A|^{k_0}} \le \frac{|A|^{k_0}}{|B|^n}\frac{\max_{x\in G}\#\{Q^B_n(x)\cap\Lambda\}}{|A|^{k_0}}$$
$$= \frac{\max_{x\in G}\#\{Q^B_n(x)\cap\Lambda\}}{|B|^n}.$$

If we take $\limsup_{n\to\infty}$ in the last inequality we have that $D^+_A(\Lambda) \le D^+_B(\Lambda)$. By using the expansiveness of B^* and using the same reasoning, we obtain that $D^+_A(\Lambda) = D^+_B(\Lambda)$. A similar argument proves that $D^-_A(\Lambda) = D^-_B(\Lambda)$.

□

As a consequence of the above result and in order to keep the exposition as clear as possible, we choose to omit the subscript A in the density and simply write $D^+(\Lambda)$, $D^-(\Lambda)$ and $D(\Lambda)$.

The next lemma shows a characterization of the subsets Λ whose upper density is finite and provides a valid version of [6, Lemma 2.3] in this context. Our proof is based on the group structure. In [24, Theorem 3.7], the authors proved the same result based on [24, Lemma 2.3], which is not satisfied in our case. (See Example 3 and [24, Lemma 2.3, item (ii)]).

Lemma 3 *Let G be an LCA group, $H \subsetneq G$ a compact open subgroup and $A \in \mathrm{Aut}(G)$ expansive with respect to H. If $\Lambda \subset G$, then the following conditions are equivalent:*

- *(i)* $D^+(\Lambda) < \infty$;
- *(ii) For some $n \in \mathbb{Z}$ there exists $N_n > 0$ such that $\#\{\Lambda \cap Q_n(A^n c)\} \le N_n$ for all $c \in C$, where C is a section for G/H.*

Additionally, if (ii) holds for some $n \in \mathbb{Z}$, it holds for every $n \in \mathbb{Z}$.

Proof $(i) \Rightarrow (ii)$. It is obvious from the definition of $D^+(\Lambda)$.

$(ii) \Rightarrow (i)$. Let $n \in \mathbb{Z}$, $C = C_0 + C_1$ as in Remark 1 and $N_n > 0$ be such that $\#\{\Lambda \cap Q_n(A^n c)\} \le N_n$ for all $c \in C$ and label for every $c \in C$, $\Lambda \cap Q_n(A^n c) = \{\lambda_{1,c}, \cdots, \lambda_{r,c}\}$ with $r = r(c) \le N_n$. Now, for $j \in \{1, \cdots, N_n\}$ and $c_0 \in C_0$ set $\Lambda_{j,c_0} := \{\lambda_{j,c_0+c_1} : c_1 \in C_1\}$. By construction, each element of Λ_{j,c_0} lies in a different coset of $G/A^n H$. Then, since there are at most $|A|N_n$ many Λ_{j,c_0} sets, we have that any coset of $G/A^n H$ contains at most $|A|N_n$ elements of Λ.

Now, since

$$A^{n+1}H = A^n(AH) = \bigcup_{c_0\in C_0} (A^n H + A^n c_0)$$

and

$$
\begin{aligned}
G = A^n G &= \bigcup_{c_0 \in C_0, c_1 \in C_1} (A^n H + A^n c_0 + A^n c_1) \\
&= \bigcup_{c_1 \in C_1} \left(\bigcup_{c_0 \in C_0} (A^n H + A^n c_0) + A^n c_1 \right),
\end{aligned}
$$

each coset of $G/(A^{n+1}H)$ has at most $|A|^2 N_n$ elements of Λ. Continuing by induction, we see that if $m > n$, then each coset of $G/(A^m H)$ has at most $|A|^{m-n+1} N_n$ elements of Λ.

As a consequence,

$$
\begin{aligned}
D^+(\Lambda) &= \limsup_{m \to +\infty} \frac{1}{|A|^m} \max_{x \in G} \#\{\Lambda \cap Q_m(x)\} \\
&\leq \limsup_{m \to +\infty} \frac{1}{|A|^m} |A|^{m-n+1} N_n \\
&= \frac{N_n}{|A|^{n-1}} < \infty.
\end{aligned}
$$

Suppose now that (ii) fails for some $n \in \mathbb{Z}$. Then, $\max_{x \in G} \#\{\Lambda \cap Q_m(x)\} = +\infty$ for every $m \geq n$. As a consequence, $D^+(\Lambda) = +\infty$. This completes the proof. □

Remark 2 Note that if $\Lambda \subset G$ and $D^+(\Lambda) < \infty$, we can deduce from the proof of Lemma 3 that for every fixed $n \in \mathbb{Z}$, $\Lambda = \bigcup_{j \in \{1, \cdots, N_n\}, c_0 \in C_0} \Lambda_{j,c_o}$ where the union is disjoint. Moreover, every set Λ_{j,c_o} is a uniformly separated set because we saw that each element of Λ_{j,c_o} lies in a different coset of $G/A^n H$ and then $\#\{\Lambda_{j,c_o} \cap [A^n H + A^n c]\} \leq 1$ for every $c \in C$. Therefore, when $D^+(\Lambda) < \infty$ we have that Λ must be a finite union of uniformly separated sets; that is, Λ must be A-separated. Since uniformly separated sets must be countable, Λ must be countable as well.

Example 4 Let G be an LCA group, $H \subseteq G$ a compact open subgroup and C a section of the quotient G/H. Consider $\Lambda := C$. Then, we have that for all $c \in C$, $\#\{C \cap Q_0(c)\} = 1$ and proceeding as in the proof of Lemma 3 we get $\#\{C \cap Q_n(A^n c)\} = |A|^n$ for all $n \in \mathbb{Z}$. Therefore, $D^+(C) = D^-(C) = 1 = D(C)$.

4 Modulation Spaces on Locally Compact Groups

In this section, we consider G a locally compact group (non necessarily abelian), with a compact open subgroup H. Using definitions and lemmas from the seminal papers [8–12], we will prove that the generalized wavelet transform continuously maps modulation spaces into Wiener spaces.

Let G be a locally compact group with right Haar measure m_G. To emphasize that G need not be abelian, in this section we will use multiplicative notation for the operation. Given $\mathcal{H}$ a Hilbert space, a unitary representation of G on $\mathcal{H}$ is a continuous homomorphism $\pi : G \to \mathcal{U}(\mathcal{H})$, where $\mathcal{U}(\mathcal{H})$ is the space of *unitary operators* on $\mathcal{H}$. For $f, g \in \mathcal{H}$ we consider $\mathcal{V}_g f : G \to \mathbb{C}$, the *generalized wavelet transform associated to the unitary representation* π (also called *voice transform* or *representation coefficients*) of f with respect to the window g defined by

$$\mathcal{V}_g f(x) := \langle f, \pi(x)g \rangle, \quad x \in G.$$

The set of *analyzing vectors* on G is given by

$$\mathcal{A} := \{g \in \mathcal{H} : \mathcal{V}_g g \in L^1(G)\}. \tag{4}$$

Note that for $g \in \mathcal{H}$ and $x \in G$, since π is a representation, $\mathcal{V}_{\pi(x)g}(\pi(x)g)(y) = \mathcal{V}_g g(x^{-1}yx)$ for all $y \in G$. Thus, $\mathcal{A}$ is invariant under π. Then, when π is an irreducible unitary representation, that is, without proper invariant subspaces, $\mathcal{A}$ must be a dense linear subspace of $\mathcal{H}$.

In the remainder of this section, we assume that G has a compact open subgroup that we will always refer to as H. Examples of non-abelian groups with that property follow.

Example 5 Fix p prime and $n \geq 2$. Then the general linear group $\mathrm{GL}_n(\mathbb{Q}_p)$ is a non-abelian locally compact group with well-understood representation theory, and $\mathrm{GL}_n(\mathbb{Z}_p)$ is a compact open subgroup [3].

We denote the space of continuous functions on G as $C := C(G)$. For $\varphi \in C$ and every $x \in G$ we see that $\|\chi_{Hx} \cdot \varphi\|_\infty = \sup_{y \in Hx} |\varphi(y)| < \infty$, where χ_E is the characteristic function of E that takes the value 1 for $x \in E$ and 0 otherwise.

For $1 \leq p < \infty$, the Wiener space $W(C, L^p)$ is defined as

$$W(C, L^p) := \{\varphi \in C : \int_G \|\chi_{Hx} \cdot \varphi\|_\infty^p dm_G(x) < \infty\}.$$

It turns out that $W(C, L^p)$ equipped with the norm given by $\|\varphi\|_{W(C,L^p)} := \left(\int_G \|\chi_{Hx} \cdot \varphi\|_\infty^p dm_G(x)\right)^{1/p}$ is a Banach space.

Additionally, if C is a section of G/H and $1 \leq p < \infty$, let us denote by $W(C, \ell^p)$ the space

$$W(C, \ell^p) := \{\varphi \in C : \sum_{x \in C} \|\chi_{Hx} \cdot \varphi\|_\infty^p < \infty\}.$$

We note that $W(C, \ell^p)$ does not depend on the choice of the section C. Furthermore, it holds that $W(C, \ell^p)$ endowed with the norm given by $\|\varphi\|_{W(C,\ell^p)} := \left(\sum_{x \in C} \|\chi_{Hx} \cdot \varphi\|_\infty^p\right)^{1/p}$ is also a Banach space.

Moreover, it can be seen that actually $W(C, \ell^p) = W(C, L^p)$ and the norms $\|\cdot\|_{W(C,L^p)}$ and $\|\cdot\|_{W(C,\ell^p)}$ coincide as we show in the next lemma.

Lemma 4 *Let G be a locally compact group with a compact open subgroup. Then, $W(C, \ell^p) = W(C, L^p)$. Moreover, for $\varphi \in C$,*

$$\|\varphi\|_{W(C,\ell^p)} = \|\varphi\|_{W(C,L^p)}.$$

Proof Let $\varphi \in C$. Since $G = \bigcup_{c\in C} Hc$ where the union is disjoint, C is a section of G/H and $m_G(H) = 1$, we can write

$$\begin{aligned}\|\varphi\|^p_{W(C,L^p)} &= \int_G \|\chi_{Hx} \cdot \varphi\|^p_\infty dm_G(x) = \sum_{c\in C} \int_{Hc} \|\chi_{Hx} \cdot \varphi\|^p_\infty dm_G(x)\\ &= \sum_{c\in C} \int_{Hc} \|\chi_{Hc} \cdot \varphi\|^p_\infty dm_G(x) = \sum_{c\in C} \|\chi_{Hc} \cdot \varphi\|^p_\infty\\ &= \|\varphi\|^p_{W(C,\ell^p)}.\end{aligned}$$

Then $\varphi \in W(C, \ell^p)$ if and only if $\varphi \in W(C, L^p)$. □

We now consider the following subset of $\mathcal{A}$,

$$\mathcal{B} = \{g \in \mathcal{H} : \mathcal{V}_g g \in W(C, L^1)\}, \tag{5}$$

which turns out to be also invariant under π. As before, when π is irreducible, $\mathcal{B}$ is dense in $\mathcal{H}$. Moreover, when additionally G is abelian, as a consequence of [12, Lemma 7.2], $\mathcal{A} = \mathcal{B}$.

Fixing an arbitrary non-zero element $g \in \mathcal{A}$, the space $\mathcal{M}^1(G)$ is given by

$$\mathcal{M}^1(G) := \{f \in \mathcal{H} : \mathcal{V}_g f \in L^1(G)\},$$

and it is called a *modulation space*. It is a Banach space with the norm $\|f\|_{\mathcal{M}^1} := \|\mathcal{V}_g f\|_{L^1}$. The set $\mathcal{M}^1$ is independent of the choice of g; i.e., different vectors in $\mathcal{A}$ give the same space with equivalent norms (see, for instance, [11, Theorem 4.2]). Considering the topological dual of $\mathcal{M}^1$, denoted as $(\mathcal{M}^1(G))'$, and $p \in [1, \infty]$, it is said that $f \in \mathcal{M}^p(G)$, the *modulation space of order* p, if $f \in (\mathcal{M}^1(G))'$ and $\|\mathcal{V}_g f\|_{L^p(G)} < +\infty$. These spaces are Banach spaces with the norm $\|f\|_{\mathcal{M}^p} = \|\mathcal{V}_g f\|_{L^p}$, and they are also independent of the choice of the window g. For more details on these spaces we refer to [11, Section 4].

It is known [12, Theorem 8.1] that for general locally compact groups, when $g \in \mathcal{B}$, $\mathcal{V}_g$ maps the space $\mathcal{M}^p$ into $W(C, \ell^p)$. We will prove now that, when G has a compact open subgroup, $\mathcal{V}_g$ maps $\mathcal{M}^p$ into $W(C, \ell^p)$ continuously.

Proposition 1 *Let G be a locally compact group with a compact open subgroup, $\pi : G \to \mathcal{U}(\mathcal{H})$ an irreducible unitary representation of G, $g\in\mathcal{B}$, and $1\leq p\leq +\infty$.*

If $f \in \mathcal{M}^p(G)$, then $\mathcal{V}_g f \in W(C, \ell^p)$. Furthermore, there exists a constant $K > 0$ such that

$$\|\mathcal{V}_g f\|_{W(C,\ell^p)} \leq K\|f\|_{\mathcal{M}^p} \tag{6}$$

for every $f \in \mathcal{M}^p(G)$.

Proof By [12, Theorem 8.1] we know that for $f \in \mathcal{M}^p(G)$ we have $\mathcal{V}_g f \in W(C, L^p)$, and that for each set $\mathcal{X} = \{x_i\}_{i\in I}$ which is a section of G/H, the linear operator given by

$$R_{\mathcal{X}} : f \mapsto (\mathcal{V}_g f(x_i))_{i\in I}$$

maps $\mathcal{M}^p(G)$ to $\ell^p(I)$ continuously; i.e., there exists $K_{\mathcal{X}} > 0$ such that

$$\|(\mathcal{V}_g f(x_i))_{i\in I}\|_{\ell^p} \leq K_{\mathcal{X}}\|f\|_{\mathcal{M}^p} \qquad \forall f \in \mathcal{M}^p(G).$$

Consider $\mathcal{F} = \{R_{\mathcal{X}} : \mathcal{X} \text{ is a section of } G/H\}$ and fix $f \in \mathcal{M}^p(G)$. Then, we have that

$$\sup_{R_{\mathcal{X}}\in\mathcal{F}} \|R_{\mathcal{X}}(f)\|_{\ell^p} = \sup_{\mathcal{X}=\{x_i\}_{i\in I}} \|(\mathcal{V}_g f(x_i))_{i\in I}\|_{\ell^p} = \|(\mathcal{V}_g f(\tilde{x}_i))_{i\in I}\|_{\ell^p}$$

where $\mathcal{X}$ is a section of G/H and the second supremum is taken over all the sections of it. Here, in the last equality, for each $i \in I$, $\tilde{x}_i \in H + x_i$ is a point that maximizes $|\mathcal{V}_g f|$ on $H + x_i$ which exists since $\mathcal{V}_g f$ is continuous and $H + x_i$ is a compact set. Moreover, $\widetilde{\mathcal{X}} = \{\tilde{x}_i\}_{i\in I}$ is a section of G/H. Then, by the uniform boundedness principle, there exists $K > 0$ such that

$$\sup_{R_{\mathcal{X}}\in\mathcal{F}} \|R_{\mathcal{X}}\| = \sup_{R_{\mathcal{X}}\in\mathcal{F}} K_{\mathcal{X}} \leq K.$$

Consequently, we have that for all sections $\{x_i\}_{i\in I}$ of G/H the following inequality holds

$$\|(\mathcal{V}_g f(x_i))_{i\in I}\|_{\ell^p} \leq K\|f\|_{\mathcal{M}^p},$$

for every $f \in \mathcal{M}^p(G)$. Now, note that, for each $f \in \mathcal{M}^p(G)$, $\|\mathcal{V}_g f\|_{W(C,\ell^p)} = \|(\mathcal{V}_g f(\tilde{x}_i))_{i\in I}\|_{\ell^p}$ for an appropriate section $\widetilde{\mathcal{X}} = \{\tilde{x}_i\}_{i\in I}$. Thus,

$$\|\mathcal{V}_g f\|_{W(C,\ell^p)} \leq K\|f\|_{\mathcal{M}^p} \qquad \forall f \in \mathcal{M}^p(G).$$

□

We shall see now that there is a valid version of Proposition 1 for projective representations.

For this, let G be a locally compact group and recall that a *projective representation* is a continuous mapping $\Pi : G \to \mathcal{U}(\mathcal{H})$ for which there exists a continuous function $\kappa : G \times G \to \mathbb{T}$, called a *2-cocycle*, such that $\Pi(x)\Pi(y) = \kappa(x, y)\Pi(xy)$. It is usual to call Π an *κ-projective representation* to emphasize the dependence of Π on κ. As we did for unitary representations, we define the *generalized wavelet transform* corresponding to a projective representation as

$$\mathcal{V}_g^{\Pi} f(x) := \langle f, \Pi(x)g \rangle,$$

for $f, g \in \mathcal{H}$.

Every projective representation of G induces a unitary representation on the *Mackey group* associated to G. If G is a locally compact group and κ is a 2-cocycle, as a topological space, the Mackey group is $G \times \mathbb{T}$, where the product is given by

$$(x_1, \tau_1)(x_2, \tau_2) = (x_1x_2, \tau_1\tau_2\kappa(x_1, x_2)),$$

for $x_1, x_2 \in G$ and $\tau_1, \tau_2 \in \mathbb{T}$. The Mackey group associated to G is a locally compact group and its Haar measure is given by the product of the Haar measures on G and $\mathbb{T}$. Then, for a κ-projective representation $\Pi : G \to \mathcal{U}(\mathcal{H})$, define $\pi : G \times \mathbb{T} \to \mathcal{U}(\mathcal{H})$ as

$$\pi(x, \tau) := \tau\Pi(x), \tag{7}$$

for $\tau \in \mathbb{T}, x \in G$. This mapping π turns out to be a unitary representation, and it is irreducible when Π is. Note that for every $f, g \in \mathcal{H}$, $x \in G$ and $\tau \in \mathbb{T}$ we have

$$\mathcal{V}_g^{\Pi} f(x) = \tau\langle f, \tau\Pi(x)g \rangle = \tau\langle f, \pi(x, \tau)g \rangle = \tau\mathcal{V}_g f(x, \tau),$$

with $\mathcal{V}_g f$ being the generalized wavelet transform associated to π. As a consequence, the sets $\mathcal{A}$ and $\mathcal{B}$ given in (4) and (5) respectively, remain equal if we use the generalized wavelet transform induced by Π instead of the one induced by the unitary representation given by (7). Then, the same holds for modulation spaces.

Therefore, we obtain the corresponding version of Proposition 1 for projective representations. This result is a particular case of [16, Theorem 2.5], with a significant distinction being our demonstration that the inclusion given by the wavelet transform is continuous.

Theorem 1 *Let G be a locally compact group with a compact open subgroup, $\Pi : G \to \mathcal{U}(\mathcal{H})$ a irreducible projective representation of G, $g \in \mathcal{B}$ and $1 \leq p \leq +\infty$. If $f \in \mathcal{M}^p(G)$ then $\mathcal{V}_g^{\Pi} f \in W(C, \ell^p)$. Furthermore, there exists a constant $K > 0$ such that*

$$\|\mathcal{V}_g^{\Pi} f\|_{W(C,\ell^p)} \leq K\|f\|_{\mathcal{M}^p} \tag{8}$$

for every $f \in \mathcal{M}^p(G)$.

Proof Note that if $H \subseteq G$ is a compact open subgroup of G, then $H \times \mathbb{T}$ is a compact open subgroup of the Mackey group of G. Also note that if C is a section of G/H, then $C \times \{1\}$ is a section of $(G \times \mathbb{T})/(H \times \mathbb{T})$. Therefore, for each $c \in C$ we have

$$\|\chi_{Hc} \cdot \mathcal{V}_g^{\Pi} f\|_\infty = \|\chi_{Hc\times\mathbb{T}} \cdot \mathcal{V}_g f\|_\infty,$$

and then

$$\|\mathcal{V}_g^{\Pi} f\|_{W(C,\ell^p)} = \|\mathcal{V}_g f\|_{W(C(G\times\mathbb{T}),\ell^p)}.$$

The result follows by applying Proposition 1. □

5 Reproducing Properties of Gabor Systems

In this section we study Bessel and frame conditions on Gabor systems in terms of density. To be precise, let us fix G, an LCA group. For every $x \in G$, the *translation operator* by x of a function $f \in L^2(G)$ is given by

$$T_x f(y) = f(y - x), \quad \text{for } m_G\text{-a.e. } y \in G.$$

For $\xi \in \widehat{G}$, the *modulation operator* by ξ of a function $f \in L^2(G)$ is defined by

$$M_\xi f(y) = \langle y, \xi \rangle f(y), \quad \text{for } m_G\text{-a.e. } y \in G.$$

Now, given a function $\varphi \in L^2(G)$ and a set $\Lambda \subseteq G \times \widehat{G}$, we define the *Gabor system generated by* φ *and* Λ as

$$S(\varphi, \Lambda) = \{M_\xi T_x \varphi\}_{(x,\xi)\in\Lambda}.$$

In order to establish frame conditions on $S(\varphi, \Lambda)$ in terms of density of Λ, we assume that G has a compact open subgroup H, and we fix $A \in \mathrm{Aut}(G)$ expansive with respect to H such that A^* is expansive with respect to $H^\perp$. Then, the densities of $\Lambda \subseteq G \times \widehat{G}$ are

$$D^+(\Lambda) := \limsup_{n\to+\infty} \frac{1}{|A|^{2n}} \max_{(x,\xi)\in G\times\widehat{G}} \#\{\Lambda \cap Q_n(x,\xi)\},$$

and

$$D^-(\Lambda) := \liminf_{n\to+\infty} \frac{1}{|A|^{2n}} \min_{(x,\xi)\in G\times\widehat{G}} \#\{\Lambda \cap Q_n(x,\xi)\},$$

where $Q_n(x, \xi)$ is defined as in (2).

On the other hand, note that translation and modulation operators are unitary in $L^2(G)$, and they satisfy the intertwining relationship $M_\xi T_x f = \langle x, \xi \rangle T_x M_\xi f$ for all $x \in G$, $\xi \in \widehat{G}$ and for all $f \in L^2(G)$. Thus, the Gabor representation $\Pi : G \times \widehat{G} \to \mathcal{U}(L^2(G))$ given by

$$\Pi(x, \xi) := M_\xi T_x \tag{9}$$

is an irreducible projective representation with 2-cocycle given by $\kappa((x_1, \xi_1) = (x_2, \xi_2)) = \langle x_1, \xi_2 \rangle$. Then, we can make use of the tools described in the previous section. In particular, we have defined the well-know *short-time Fourier transform*

$$V_g f(x, \xi) := \langle f, M_\xi T_x g \rangle = \mathcal{V}_g^\Pi f(x, \xi),$$

where, $f, g \in L^2(G)$, and $(x, \xi) \in G \times \widehat{G}$. For fixed $f, g \in L^2(G)$, $V_g f$ is well-defined and continuous on $G \times \widehat{G}$.

When we consider translations along a section of G/H and modulations along a section of $\widehat{G}/H^\perp$ of the function χ_H, it turns out that the obtained Gabor system is an orthonormal basis for $L^2(G)$. See [16, Theorem 2.7, Case II] for a proof of this fact.

Lemma 5 *Let $C \subset G$ and $D \subset \widehat{G}$ be sections of G/H and $\widehat{G}/H^\perp$ respectively, and consider $\Lambda = C \times D$. Then, $S(\chi_H, \Lambda)$ is an orthonormal basis for $L^2(G)$.*

5.1 Bessel Sequences

In this section, we show one necessary and one sufficient condition for a Gabor system to be a Bessel sequence of $L^2(G)$.

We begin by proving that if a Gabor system is a Bessel sequence, then Λ must have finite upper density. This was proved before for $\mathbb{R}^d$ in [6, Theorem 3.1].

Theorem 2 *Let $\varphi \in L^2(G)$ and let $\Lambda \subseteq G \times \widehat{G}$. If $S(\varphi, \Lambda)$ is a Bessel sequence in $L^2(G)$, then $D^+(\Lambda) < \infty$.*

Proof Let $f \in L^2(G)$ with $\|f\|_2 = 1$ such that $\langle \varphi, f \rangle \neq 0$ and define $A_\varphi f : G \times \widehat{G} \to \mathbb{R}_{\geq 0}$

$$A_\varphi f(x, \xi) := |V_\varphi f(x, \xi)|.$$

Then, $A_\varphi f$ is continuous on $G \times \widehat{G}$.

As $A_\varphi f \neq 0$, there exists $(x_0, \xi_0) \in G \times \widehat{G}$ and $n_0 \in \mathbb{Z}$ such that $\eta := \inf\{A_\varphi f(x, \xi) : (x, \xi) \in Q_{n_0}(x_0, \xi_0)\} > 0$.

If we had $D^+(\Lambda) = \infty$, by Lemma 3, for each $N > 0$ there should exist some (x_N, ξ_N) such that $\#\{\Lambda \cap Q_{n_0}(x_N, \xi_N)\} \geq N$.

Now, note that if $(x, \xi) \in Q_{n_0}(x_N, \xi_N)$, then $(x, \xi) - (x_N, \xi_N) + (x_0, \xi_0) \in Q_{n_0}(x_0, \xi_0)$ and thus

$$|\langle f, M_{\xi - \xi_N + \xi_0} T_{x - x_N + x_0} \varphi \rangle| \geq \eta.$$

Since $|\langle f, M_{\xi - \xi_N + \xi_0} T_{x - x_N + x_0} \varphi \rangle| = |\langle M_{\xi_0 - \xi_N} T_{x_0 - x_N} f, M_\xi T_x \varphi \rangle|$ we have that

$$\sum_{(x,\xi) \in \Lambda \cap Q_{n_0}(x_N, \xi_N)} |\langle M_{\xi_0 - \xi_N} T_{x_0 - x_N} f, M_\xi T_x \varphi \rangle|^2 \geq \eta^2 N,$$

for all $N \in \mathbb{N}$. Hence, $S(\varphi, \Lambda)$ can not be a Bessel sequence because $\|M_{\xi_0 - \xi_N} T_{x_0 - x_N} f\|_2 = \|f\|_2 = 1$. □

The above theorem extends to a finite union of Gabor systems. More precisely, let $\Lambda_1, \ldots, \Lambda_r \subseteq G \times \widehat{G}$ be subsets each indexed in $I_1, \ldots, I_r$, respectively. That is, $\Lambda_k = \{(x_{i,k}, \xi_{i,k})\}_{i \in I_k}$, for $1 \leq k \leq r$. Define $I = \{(i, k) : i \in I_k, 1 \leq k \leq r\}$ and Λ as the set $\{(x_{i,k}, \xi_{i,k}) : (i, k) \in I\}$. By abuse of notation, we will simply write $\Lambda = \bigcup_{k=1}^r \Lambda_k$ and say that Λ is the *disjoint union of* $\Lambda_1, \ldots, \Lambda_r$. Once this is clear, we can state the desired result.

Corollary 1 *For $1 \leq k \leq r$, let $\varphi_k \in L^2(G)$ and $\Lambda_k \subseteq G \times \widehat{G}$. Consider Λ the disjoint union of $\Lambda_1, \ldots, \Lambda_r$. If $\bigcup_{k=1}^r S(\varphi_k, \Lambda_k)$ is a Bessel sequence, then $D^+(\Lambda) < \infty$.*

Proof Note that since $\bigcup_{k=1}^r S(\varphi_k, \Lambda_k)$ is a Bessel sequence, so is $S(\varphi_k, \Lambda_k)$ for every $1 \leq k \leq r$. Then, by Theorem 2 we have that $D^+(\Lambda_k) < \infty$ for every $1 \leq k \leq r$.

Now, since Λ is the disjoint union of $\Lambda_1, \ldots, \Lambda_r$, for each $n \in \mathbb{Z}$ and $(x, \xi) \in G \times \widehat{G}$,

$$\#\{\Lambda \cap Q_n(x, \xi)\} = \sum_{k=1}^r \#\{\Lambda_k \cap Q_n(x, \xi)\},$$

and as a consequence

$$\sum_{k=1}^r D^-(\Lambda_k) \leq D^-(\Lambda) \leq D^+(\Lambda) \leq \sum_{k=1}^r D^+(\Lambda_k).$$

From here the conclusion follows. □

In what follows we shall prove a weaker converse of Theorem 2. For this to be true, it is assumed that the generating function of the Gabor system has a particular decay; that is, it must be a function of $\mathscr{M}^1(G)$.

As a consequence of [22, Theorem 4.7] (see also [8, Theorem 11]) we have that, when the representation involved is the Gabor representation (9), $\mathcal{A} = \mathcal{B} = \mathcal{M}^1(G)$, and then, Theorem 1 holds for $g \in \mathcal{M}^1(G)$.

Then, we have the following result, which is a generalization of [18, Theorem 12] to our setting.

Theorem 3 *Let $\varphi \in \mathcal{M}^1(G)$, $\varphi \neq 0$ and $\Lambda \subseteq G \times \widehat{G}$ any set with $D^+(\Lambda) < +\infty$. Then $S(\varphi, \Lambda)$ is a Bessel sequence.*

Proof By Lemma 3, since $D^+(\Lambda) < +\infty$ there exists $N_0 > 0$ such that $\#\{\Lambda \cap Q_0(c,d)\} \leq N_0$ for all $(c,d) \in C \times D$. Then for $f \in L^2(G)$,

$$\begin{aligned}\sum_{(x,\xi)\in\Lambda} |\langle f, M_\xi T_x \varphi\rangle|^2 &= \sum_{(x,\xi)\in\Lambda} |V_\varphi f(x,\xi)|^2 \\ &= \sum_{(c,d)\in C\times D} \sum_{(x,\xi)\in\Lambda\cap Q_0(c,d)} |V_\varphi f(x,\xi)|^2 \\ &\leq \sum_{(c,d)\in C\times D} N_0 \sup_{(x,\xi)\in Q_0(c,d)} |V_\varphi f(x,\xi)|^2 \\ &= N_0 \|V_\varphi f\|^2_{W(C,\ell^2)} \leq N_0 K \|f\|^2_{\mathcal{M}^2},\end{aligned}$$

where the last inequality is the result of Theorem 1. Finally, since $\|f\|_{\mathcal{M}^2} = \|V_\varphi f\|_2$ and by the well-known orthogonality relationship of the short-time Fourier transform (see [11, Section 2.2]), $\|V_\varphi f\|_2 = \|f\|_2 \|\varphi\|_2$, we conclude the result. □

It is known that every locally compact abelian group G is algebraically and topologically isomorphic to $\mathbb{R}^d \times G_0$, where G_0 is an LCA group with a compact open subgroup (see, e.g., [7]). For the case where G_0 is an expansible group we can combine Theorem 3 with [18, Theorem 12] to prove a similar statement for the product group $\mathbb{R}^d \times G_0$.

Proposition 2 *Let G_0 be an expansible LCA group. Take $g_1 \in M^1(\mathbb{R}^d)$, $g_2 \in M^1(G_0)$ and $\Lambda = \Lambda_1 \times \Lambda_2$, where $\Lambda_1 \subseteq \mathbb{R}^{2d}$ and $\Lambda_2 \subseteq G_0 \times \widehat{G_0}$. Suppose that Λ_1 and Λ_2 have finite upper density, where the density of Λ_1 is the usual Beurling density defined in the introduction, and the density of Λ_2 is as in Definition 3. If we consider $g \in L^2(\mathbb{R}^d \times G_0)$ given by $g = g_1 \otimes g_2$, then $S(g, \Lambda)$ is a Bessel sequence for $L^2(\mathbb{R}^d \times G_0)$.*

Proof During this proof we use $\|\cdot\|_{W(\mathbb{R}^n)}$ and $\|\cdot\|_{W(G_0)}$ to denote the norm in the Wiener spaces $W(C, \ell^2)$ with underlying spaces $\mathbb{R}^n$ and G_0 respectively.

First observe that for $(x_1, x_2, \xi_1, \xi_2) \in (\mathbb{R}^d \times G_0) \times (\mathbb{R}^d \times \widehat{G_0})$ and $f \in L^2(\mathbb{R}^d \times G_0)$, we have

$$V_g f(x_1, x_2, \xi_1, \xi_2)$$

$$= \int_{\mathbb{R}^d} \int_{G_0} f(y_1, y_2) \overline{g(y_1 - x_1, y_2 - x_2) \langle y_1, \xi_1 \rangle \langle y_2, \xi_2 \rangle} dm_{G_0}(y_2) dy_1$$

$$= \int_{\mathbb{R}^d} \int_{G_0} f(y_1, y_2) \overline{g_1(y_1 - x_1) g_2(y_2 - x_2) \langle y_1, \xi_1 \rangle \langle y_2, \xi_2 \rangle} dm_{G_0}(y_2) dy_1$$

$$\times \int_{\mathbb{R}^d} \overline{g_1(y_1 - x_1) \langle y_1, \xi_1 \rangle} \left(\int_{G_0} f(y_1, y_2) \overline{g_2(y_2) \langle y_2, \xi_2 \rangle} dm_{G_0}(y_2) \right) dy_1$$

$$= V_{g_1} \left(V_{g_2} f_{\bullet}(x_2, \xi_2) \right) (x_1, \xi_1),$$

where by $f_{\bullet}$ we denote the function the is obtained when the first variable of f is fixed.

On the other hand, since both upper densities are finite, we know from Lemma 3 and [6, Lemma 2.3] that there exist N_0 and $N_1 \in \mathbb{N}$ such that

$$\#(\Lambda_2 \cap Q_0(x, \xi)) \leq N_0 \text{ for all } (x, \xi) \in G_0 \times \widehat{G_0} \text{ and}$$

$$\#(\Lambda_1 \cap B_1(t)) \leq N_1 \text{ for all } t \in \mathbb{R}^{2d},$$

where $B_1(0) = [0, 1]^{2d}$ and $B_1(t) = B_1(0) + t$.

Putting this together we obtain

$$\sum_{(x_1, x_2, \xi_1, \xi_2) \in \Lambda} |V_g f(x_1, x_2, \xi_1, \xi_2)|^2$$

$$= \sum_{(x_2, \xi_2) \in \Lambda_2} \sum_{(x_1, \xi_1) \in \Lambda_1} |V_{g_1} \left(V_{g_2} f_{\bullet}(x_2, \xi_2) \right) (x_1, \xi_1)|^2$$

$$= \sum_{(x_2, \xi_2) \in \Lambda_2} \sum_{j \in \mathbb{Z}^{2d}} \sum_{(x_1, \xi_1) \in \Lambda_1 \cap B_1(j)} |V_{g_1} \left(V_{g_2} f_{\bullet}(x_2, \xi_2) \right) (x_1, \xi_1)|^2$$

$$\leq \sum_{(x_2, \xi_2) \in \Lambda_2} \sum_{j \in \mathbb{Z}^{2d}} N_1 \sup_{(x_1, \xi_1) \in \Lambda_1 \cap B_1(j)} |V_{g_1} \left(V_{g_2} f_{\bullet}(x_2, \xi_2) \right) (x_1, \xi_1)|^2$$

$$= N_1 \sum_{(x_2, \xi_2) \in \Lambda_2} \| V_{g_1} (V_{g_2} f_{\bullet}(x_2, \xi_2)) \|^2_{W(\mathbb{R}^d)}$$

$$\leq N_1 \sum_{(x_2, \xi_2) \in \Lambda_2} \| V_{g_2} f_{\bullet}(x_2, \xi_2) \|^2_{L^2(\mathbb{R}^d)} K_1 \| g_1 \|^2_{L^2(\mathbb{R}^d)},$$

where in the last inequality we used [18, Theorem 12] and the fact that $V_{g_2} f_{\bullet}(x_2, \xi_2) \in L^2(\mathbb{R}^d)$ for a.e. $(x_2, \xi_2) \in G_0 \times \widehat{G_0}$. Now, using Fubini and Theorem 1 we have

$$N_1 K_1 \| g_1 \|^2_{L^2(\mathbb{R}^d)} \sum_{(x_2, \xi_2) \in \Lambda_2} \| V_{g_2} f_{\bullet}(x_2, \xi_2) \|^2_{L^2(\mathbb{R}^d)}$$

$$
\begin{aligned}
&= K_0 \sum_{(x_2,\xi_2)\in\Lambda_2} \int_{\mathbb{R}^d} |V_{g_2} f_{y_1}(x_2,\xi_2)|^2 dy_1 \\
&= K_0 \int_{\mathbb{R}^d} \sum_{(c,d)\in C\times D} \sum_{(x_2,\xi_2)\in\Lambda_2\cap Q_0(c,d)} |V_{g_2} f_{y_1}(x_2,\xi_2)|^2 dy_1 \\
&\le K_0 \int_{\mathbb{R}^d} \sum_{(c,d)\in C\times D} N_0 \sup_{(x_2,\xi_2)\in\Lambda_2\cap Q_0(c,d)} |V_{g_2} f_{y_1}(x_2,\xi_2)|^2 dy_1 \\
&= K_0 N_0 \int_{\mathbb{R}^d} \|V_{g_2} f_{y_1}\|^2_{W(G_0)} dy_1 \\
&\le K \int_{\mathbb{R}^d} \|f_{y_1}\|^2_{L^2(G_0)} dy_1 \\
&= K \int_{\mathbb{R}^d} \int_{G_0} |f(y_1,y_2)|^2 dy_2 dy_1 = K\|f\|^2_{L^2(\mathbb{R}^d\times G_0)}.
\end{aligned}
$$

Since

$$
\sum\nolimits_{(x_1,x_2,\xi_1,\xi_2)\in\Lambda} |\langle f, M_{(\xi_1,\xi_2)} T_{(x_1,x_2)} g\rangle|^2 = \sum\nolimits_{(x_1,x_2,\xi_1,\xi_2)\in\Lambda} |V_g f(x_1,x_2,\xi_1,\xi_2)|^2,
$$

the above computation shows that $S(g,\Lambda)$ is a Bessel sequence. □

As in the Euclidean case (see for instance [6, Theorem 1.1] and [19, Theorem 5.7]), it is possible to prove a Density Theorem for Gabor frames in this context as well. This is a consequence of [4, Corollary 3.4] which has a very sophisticated proof. However, in our setting, we can easily deduce it from [14, Theorem 4]. We include the reasoning for the sake of completeness.

Theorem 4 *Let $\Lambda, \Delta \subseteq G\times\widehat{G}$ be uniformly separated sets and $\varphi, h \in L^2(G)$ such that $S(\varphi,\Lambda)$ and $S(h,\Delta)$ are a frame and a Riesz basis for $L^2(G)$ respectively. Then*

$$
D^-(\Delta) \le D^-(\Lambda) \quad \textit{and} \quad D^+(\Delta) \le D^+(\Lambda).
$$

Proof First, we will modify $h \in L^2(G)$ to be the scalar multiple $\tilde{h}$ such that $\|\tilde{h}\|_2^2 = \beta$, where β is the upper frame bound of $S(\varphi,\Lambda)$. With this replacement $S(\tilde{h},\Delta)$ is still a Riesz basis for $L^2(G)$. By [14, Theorem 4] for a given $\varepsilon > 0$ there exists a compact set $L \subseteq G\times\widehat{G}$ such that

$$
(1-\varepsilon)\#\{\Delta\cap(K+(x,\xi))\} \le \#\{\Lambda\cap(K+L+(x,\xi))\}, \tag{10}
$$

for every compact set $K \subseteq G\times\widehat{G}$ and for every $(x,\xi)\in G\times\widehat{G}$.

Since both G and $\widehat{G}$ are expansible and L is compact, there exists $n_0 \in \mathbb{N}$ such that $Q_n(0,0) + L = Q_n(0,0)$ for all $n \ge n_0$. Then, (10) implies that

$$
(1-\varepsilon)\#\{\Delta\cap Q_n(x,\xi)\} \le \#\{\Lambda\cap Q_n(x,\xi)\}
$$

for all $(x, \xi) \in G \times \widehat{G}$ and $n \geq n_0$. Dividing by $|A|^{2n}$ and taking $\limsup$ (or respectively $\liminf$) we obtain

$$(1-\varepsilon)D^{\pm}(\Delta) \leq D^{\pm}(\Lambda)$$

for every $\varepsilon > 0$. From here the result follows. □

With the above comparison result, it is possible to deduce a density result.

Corollary 2 *Let $\Lambda \subset G \times \widehat{G}$ be a uniformly separated set and $\varphi \in L^2(G)$.*

(i) If $S(\varphi, \Lambda)$ is a frame for $L^2(G)$ then $D^-(\Lambda) \geq 1$;
(ii) If $S(\varphi, \Lambda)$ is a Riesz basis for $L^2(G)$ then $D(\Lambda) = 1$.

Proof By Lemma 5, the system $S(\chi_H, \Delta)$ with Δ being a section of $G/H \times \widehat{G}/H^{\perp}$ is an orthonormal basis for $L^2(G)$ and, as we calculate in Example 4, $D(\Delta) = 1$. Let us now consider Λ and φ as in the statement. If $S(\varphi, \Lambda)$ is a frame for $L^2(G)$ then by Theorem 4 we have $D^-(\Lambda) \geq D^-(\Delta) = 1$. If, on the other hand, $S(\varphi, \Lambda)$ is a Riesz basis for $L^2(G)$ then we apply Theorem 4 twice interchanging the role of each system to obtain $D(\Lambda) = 1$. □

We finish by showing that in our setting, we can construct Gabor frames generated by sets with density $|A|^{2k}$ for every $k \geq 0$.

Example 6 Let G be an expansible group and A an expansive automorphism with respect to H, a compact open subgroup of G. Let $k \geq 0$ and consider $\varphi := |A|^{k/2}\chi_{A^{-k}H}$ and $\Lambda = \Delta \times \Gamma$ with Δ a section of $G/A^{-k}H$ and Γ a section of $\widehat{G}/A^{*-k}H^{\perp}$. Then we can write $\Lambda = \Lambda_1 + \Lambda_2$ where Λ_1 is a section of $G/H \times \widehat{G}/H^{\perp}$ and Λ_2 is a section of $H/A^{-k}H \times H^{\perp}/A^{*-k}H^{\perp}$.

We want to calculate $D(\Lambda)$. In that sense, for $n \in \mathbb{N}$ we have

$$\begin{aligned}
\max_{(x,\xi)\in G\times\widehat{G}} \#(Q_n(x,\xi)\cap\Lambda) &\leq \sum_{\lambda_2\in\Lambda_2} \max_{(x,\xi)\in G\times\widehat{G}} \#(Q_n(x,\xi)\cap(\Lambda_1+\lambda_2)) \\
&= \sum_{\lambda_2\in\Lambda_2} \max_{(x,\xi)\in G\times\widehat{G}} \#(Q_n(x,\xi)\cap\Lambda_1) \\
&= \#\Lambda_2 \max_{(x,\xi)\in G\times\widehat{G}} \#(Q_n(x,\xi)\cap\Lambda_1) \\
&= |A|^{2k} \max_{(x,\xi)\in G\times\widehat{G}} \#(Q_n(x,\xi)\cap\Lambda_1).
\end{aligned}$$

Dividing by $|A|^{2n}$, taking $\limsup$ and applying Lemma 4 we obtain

$$D^+(\Lambda) \leq |A|^{2k} D^+(\Lambda_1) = |A|^{2k}.$$

By doing an analogous calculation with the minimum, we obtain $D^-(\Lambda) \geq |A|^{2k} D^-(\Lambda_1) = |A|^{2k}$ and therefore $D(\Lambda) = |A|^{2k}$.

On the other hand, we can also write $\Gamma = \Gamma_1 + \Gamma_2$ where Γ_1 and Γ_2 are sections of $\widehat{G}/A^{*k}H^{\perp}$ and $A^{*k}H^{\perp}/A^{*-k}H^{\perp}$ respectively. As a consequence, by Lemma 5, for every $\gamma_2 \in \Gamma_2$, the system $S(\varphi, \Delta \times (\Gamma_1 + \gamma_2))$ is an orthonormal basis for $L^2(G)$. Then, $S(\varphi, \Lambda)$ is a tight frame with bound $\#\Gamma_2 = |A|^{2k}$.

Lemma 6 *Let G, A and H as in the prior example. Consider $\varphi = \chi_H$, and for $k \in \mathbb{Z}$ let Λ_k be a section of $G/H \times \widehat{G}/A^{*-k}H^{\perp}$. Then the following holds*

(i) if $k = 0$ then $S(\varphi, \Lambda_0)$ is an orthonormal basis for $L^2(G)$,
(ii) if $k > 0$ then $S(\varphi, \Lambda_k)$ is a tight frame for $L^2(G)$ with constant $|A|^k$,
(iii) and if $k < 0$ then $S(\varphi, \Lambda_k)$ is an incomplete system for $L^2(G)$.

Proof When $k = 0$ the set Λ_0 is a section of $G/H \times \widehat{G}/H^{\perp}$ and by Lemma 5 it is an orthonormal basis. If $k > 0$ the same procedure as in the prior example can be used to prove that $S(\varphi, \Lambda_k)$ is a tight frame with constant $|A|^k$. For the remaining case, let us remember $H^{\perp} \subsetneq A^{*-k}H^{\perp}$ and choose Γ to be a section of $G/H \times \widehat{G}/H^{\perp}$ containing Λ_k properly. Since by Lemma 5 the system $S(\varphi, \Gamma)$ is an orthonormal basis for $L^2(G)$ and $S(\varphi, \Lambda) \subsetneq S(\varphi, \Gamma)$, the statement is proven. □

Acknowledgments This research was supported by grants: V.P. and R.N. were supported by PICT 2018-3399 (ANPCyT), PICT 2019-03968 (ANPCyT) and CONICET PIP 11220210100087.

We would like to thank the referees for their comments, since they have helped us to improve the manuscript.

References

1. Agora, E., Antezana, J., Cabrelli, C.: Multi-tiling sets, Riesz bases, and sampling near the critical density in LCA groups. Adv. Math. **285**, 454–477 (2015)
2. Benedetto, J., Benedetto, R.: A wavelet theory for local fields and related groups. J. Geom. Anal. **14**(3), 423–456 (2004)
3. Bushnell, C., Henniart, G.: The Local Langlands Conjecture for GL(2), volume 335 of Grundlehren der mathematischen Wissenschaften [Fundamental Principles of Mathematical Sciences]. Springer, Berlin (2006)
4. Caspers, M., van Velthoven, J.T.: Overcompleteness of coherent frames for unimodular amenable groups. Ark. Mat. **61**(2), 277–299 (2023)
5. Christensen, O.: An Introduction to Frames and Riesz Bases. Applied and Numerical Harmonic Analysis, 2nd edn. Birkhäuser/Springer, Cham (2016)
6. Christensen, O., Deng, B., Heil, C.: Density of Gabor frames. Appl. Comput. Harmon. Anal. **7**(3), 292–304 (1999)
7. Deitmar, A., Echterhoff, S.: Principles of Harmonic Analysis. Universitext, 2nd edn. Springer, Cham (2014)
8. Feichtinger, H.G.: On a new Segal algebra. Monatsh. Math. **92**, 269–289 (1981)
9. Feichtinger, H.G.: Modulation spaces on locally compact Abelian groups. Technical report, University of Vienna, January 1983
10. Feichtinger, H.G.: Modulation spaces on locally compact Abelian groups. In: Radha, R., Krishna, M., Thangavelu, S. (eds.) Proc. Internat. Conf. on Wavelets and Applications, pp. 1–56. NuHAG, New Delhi Allied Publishers (2003)

11. Feichtinger, H., Gröchenig, K.: Banach spaces related to integrable group representations and their atomic decompositions. I. J. Funct. Anal. **86**(2), 307–340 (1989)
12. Feichtinger, H., Gröchenig, K.: Banach spaces related to integrable group representations and their atomic decompositions. II. Monatsh. Math. **108**(2–3), 129–148 (1989)
13. Gröchenig, K.: Foundations of Time-frequency Analysis. Applied and Numerical Harmonic Analysis. Birkhäuser Boston, Boston (2001)
14. Gröchenig, K.: The homogeneous approximation property and the comparison theorem for coherent frames. Sampl. Theory Signal Image Process. **7**(3), 271–279 (2008)
15. Gröchenig, K., Heil, C.: Modulation spaces and pseudodifferential operators. Integral Equ. Oper. Theory **34**(4), 439–457 (1999)
16. Gröchenig, K., Strohmer, T.: Pseudodifferential operators on locally compact abelian groups and Sjöstrand's symbol class. J. Reine Angew. Math. **613**, 121–146 (2007)
17. Gröchenig, K., Kutyniok, G., Seip, K.: Landau's necessary density conditions for LCA groups. J. Funct. Anal. **255**(7), 1831–1850 (2008)
18. Heil, C.: History and evolution of the density theorem for Gabor frames. J. Fourier Anal. Appl. **13**(2), 113–166 (2007)
19. Heil, C.: The density theorem and the homogeneous approximation property for Gabor frames. In: Representations, Wavelets, and Frames, Appl. Numer. Harmon. Anal., pp. 71–102. Birkhäuser Boston, Boston (2008)
20. Hewitt, E., Ross, K.: Abstract Harmonic Analysis. Vol. I: Structure of Topological Groups. Integration Theory, Group Representations. Die Grundlehren der mathematischen Wissenschaften, Band 115. Academic Press/Springer, New York/Berlin (1963)
21. Hewitt, E., Ross, K.: Abstract Harmonic Analysis. Vol. II: Structure and Analysis for Compact Groups. Analysis on Locally Compact Abelian Groups. Die Grundlehren der mathematischen Wissenschaften, Band 152. Springer, New York (1970)
22. Jakobsen, M.: On a (no longer) new Segal algebra: a review of the Feichtinger algebra. J. Fourier Anal. Appl. **24**(6), 1579–1660 (2018)
23. Landau, H.J.: Necessary density conditions for sampling and interpolation of certain entire functions. Acta Math. **117**, 37–52 (1967)
24. Seyedi, N.S., Kamyabi Gol, R.A.: Beurling density on expansible locally compact groups and systems of translations. J. Math. Anal. Appl. **490**(2), 124319, 13 (2020)
25. Strohmer, T., Heath Jr., R.: Grassmannian frames with applications to coding and communication. Appl. Comput. Harmon. Anal. **14**(3), 257–275 (2003)

Fourier Optimization and Consequences of the Generalized Riemann Hypothesis

Emily Quesada-Herrera

Abstract We give an exposition of some connections between Fourier optimization problems and problems in number theory. In particular, we present some recent conditional bounds under the generalized Riemann hypothesis, achieved via a Fourier optimization framework, on bounding the maximum possible gap between consecutive prime numbers represented by a given quadratic form; and on bounding the least quadratic non-residue modulo a prime number. This is based on joint works with Emanuel Carneiro, Andrés Chirre, Micah Milinovich, and Antonio Pedro Ramos.

Keywords Fourier optimization · Generalized Riemann hypothesis · Quadratic forms

1 Introduction

This article is an exposition of connections between Fourier analysis and number theory, and is partly based on the article [11] (joint with Andrés Chirre) and the preprint [7] (joint with Emanuel Carneiro, Micah B. Milinovich, and Antonio Pedro Ramos). We present two results, conditional on the generalized Riemann hypothesis, to illustrate this connection. Our results are inequalities regarding problems in number theory, and Fourier analysis was used to optimize the value of the explicit constants that appear in these inequalities.

To state the problems under consideration and our results, consider first quadratic and cubic residues modulo a prime number p. An integer a is a quadratic residue mod p if there exists an integer u such that $u^2 \equiv a \bmod p$. If this is not the case, a

E. Quesada-Herrera (✉)
Department of Mathematics and Computer Science, University of Lethbridge, Lethbridge, AB, Canada
e-mail: emily.quesadaherrera@uleth.ca

P. E. Harris et al. (eds.), *Women in Mathematics in Latin America*,
Research Perspectives Ghent Analysis and PDE Center 14,
https://doi.org/10.1007/978-3-031-97857-9_15

is a quadratic non-residue. Similarly, a is a cubic residue if there exists an integer u such that $u^3 \equiv a \bmod p$.

Our first problem involves the study of integers and prime numbers represented by quadratic forms. This is a classical problem in number theory, going back at least to Fermat - he realized, for instance, that a prime number p is a sum of two squares if and only if $p \equiv 1 \bmod 4$. A more subtle example is the quadratic form x^2+27y^2. A conjecture of Euler, proven by Gauss, says that a prime p has the form $p = x^2 + 27y^2$ if and only if both $p \equiv 1 \pmod 3$ and 2 is a cubic residue mod p. The following is an illustration of one of our results in [11].

Theorem 1 *Assume the generalized Riemann hypothesis. Then, for all sufficiently large t, there is always a prime number p of the form $p = x^2 + 27y^2$ between t and $t + \frac{46}{25}\sqrt{t}\log t$.*

Our second result is about the maximum possible size of the least quadratic non-residue modulo a prime number. In [7], we show the following.

Theorem 2 *Assume the generalized Riemann hypothesis. Let n_p be the least quadratic non-residue modulo a prime p. If p is sufficiently large, then*

$$n_p < \frac{61}{80} \log^2 p.$$

Below we give more information on these problems and their background.

2 Fourier Transform

The Fourier transform is certainly one of the most fundamental objects in mathematics and applied mathematics, as it is used to model a variety of oscillatory phenomena. For an integrable function $f : \mathbb{R} \to \mathbb{C}$, we may define the Fourier transform of f by

$$\widehat{f}(\xi) := \int_{\mathbb{R}} f(x)e^{-2\pi i \xi x} dx \quad (\xi \in \mathbb{R}).$$

A Fourier optimization problem is one where we impose conditions on a function and its Fourier transform, and seek to optimize a given quantity. These problems often have surprising applications, such as in the theory of the Riemann zeta-function and the distribution of its zeros [3–5, 8–10, 18], in bounding prime gaps [6, 10, 11], and in the theory of sphere packings [12, 13, 22]. A classical reference on using extremal functions in Fourier analysis is Vaaler's 1985 survey [21].

3 Riemann Zeta-Function and Dirichlet L-functions

Understanding the behavior of the Riemann zeta-function and the distribution of its zeros is a crucial problem in number theory, being related to the distribution of primes. The Riemann zeta-function is defined in the half-plane $\mathrm{Re}s > 1$ by

$$\zeta(s) = \sum_{n=1}^{\infty} \frac{1}{n^s} = \prod_p \left(1 - \frac{1}{p^s}\right)^{-1}. \tag{1}$$

The *Euler product*—the product over primes in (1)—is the starting point of the connection between this function and the distribution of primes. It is possible to show that $\zeta(s)$ extends to a meromorphic function in the entire complex plane, with a simple pole at $s = 1$, and that $\zeta(-2n) = 0$ for all positive integers n. These are called the *trivial zeros*. Furthermore, one can show that all non-trivial zeros of $\zeta(s)$ must lie on the *critical strip* $\{s \in \mathbb{C} : 0 < \mathrm{Re}s < 1\}$. The Riemann hypothesis (RH), conjectured in 1859 [19], states that all non-trivial zeros actually lie on the *critical line*, $\mathrm{Re}s = \frac{1}{2}$. For further background on $\zeta(s)$ and primes, see [14, 15, 17, 20].

Dirichlet L-functions are one generalization of $\zeta(s)$, and were originally introduced to study primes in arithmetic progressions. Instead of giving a full definition, let us see a concrete example. Consider the Legendre symbol, defined by

$$\chi(a) = \left(\frac{a}{p}\right) := \begin{cases} 0 & \text{if } p \mid a; \\ 1 & \text{if } p \nmid a \text{ and } a \text{ is a quadratic residue mod } p; \\ -1 & \text{if } p \nmid a \text{ and } a \text{ is a quadratic non-residue mod } p. \end{cases}$$

This is an example of a *Dirichlet character* modulo p, and in this case, it codifies information about quadratic non-residues. We consider the associated Dirichlet L-function, defined initially (and then analytically continued), for $\mathrm{Re}s > 1$, by

$$L(s, \chi) = \sum_{n=1}^{\infty} \frac{\chi(n)}{n^s} = \prod_p \left(1 - \frac{\chi(p)}{p^s}\right)^{-1}. \tag{2}$$

The *generalized Riemann hypothesis* (GRH) conjectures that all non-trivial zeros (i.e., those in the critical strip) of $L(s, \chi)$ also have real part $\frac{1}{2}$. See, for instance, [14, 15] for details and a full definition.

3.1 A Link Between Worlds

A bridge that connects Fourier analysis with number theory is the Guinand-Weil explicit formula, and it is an ingredient in many of the aforementioned works. It connects an arbitrary function h, its Fourier transform $\widehat{h}$, prime numbers, and zeros of the Riemann zeta-function or an L-function:

Lemma 1 (Guinand-Weil Explicit Formula) *Let $h(s)$ be analytic in the strip $|\mathrm{Im}\, s| \le \frac{1}{2} + \varepsilon$ for some $\varepsilon > 0$, and assume that $|h(s)| \le C(1+|s|)^{-(1+\delta)}$ for some $\delta > 0$ and $C > 0$. Let χ be a primitive Dirichlet character modulo q. Then*

$$\sum_{\rho_\chi} h\left(\frac{\rho_\chi - \frac{1}{2}}{i}\right) = \widehat{h}(0)\frac{\log(q/\pi)}{2\pi} + \frac{1}{2\pi}\int_{-\infty}^{\infty} h(u)\,\mathrm{Re}\,\frac{\Gamma'}{\Gamma}\left(\frac{2-\chi(-1)}{4} + \frac{iu}{2}\right) du$$
$$-\frac{1}{2\pi}\sum_{n\ge 2}\frac{\Lambda(n)}{\sqrt{n}}\left\{\chi(n)\,\widehat{h}\left(\frac{\log n}{2\pi}\right) + \overline{\chi(n)}\,\widehat{h}\left(-\frac{\log n}{2\pi}\right)\right\},$$

where the sum on the left-hand side runs over the non-trivial zeros ρ_χ of $L(s,\chi)$, and $\Lambda(n)$ is the von Mangoldt function defined to be $\log p$ if $n = p^k$, p a prime and $k \ge 1$, and zero otherwise.

Here, the information about prime numbers is codified in the Von Mangoldt function $\Lambda(n)$, and, in our case above, $\chi(n)$ contains information about quadratic residues. While this gives a bridge between the fields, formulating a suitable Fourier optimization problem is a subtle and delicate matter.

4 Prime Gaps and Quadratic Forms

Due to the connection between primes numbers and zeros of the Riemann zeta-function and other L-functions, it is useful to study the consequences of RH and GRH, as it may allow us to better understand the limits of our current techniques when tackling a certain problem. Let p_n be the n-th prime number. Assuming RH, Cramér showed that the gap between consecutive prime numbers satisfies

$$p_{n+1} - p_n \le (C + o(1))\sqrt{p_n}\log p_n \tag{3}$$

for an absolute constant $C > 0$. The order of magnitude here has not been improved in the past 90 years, and the efforts have been in reducing the explicit value of C. Carneiro, Milinovich and Soundararajan [6] developed a Fourier optimization strategy to tackle this. They introduce the following Fourier optimization problem.

Extremal Problem 1
Define $\mathcal{A}^+$ to be the class of even and continuous functions $F : \mathbb{R} \to \mathbb{R}$, with $F \in L^1(\mathbb{R}) \setminus \{0\}$. For a parameter $1 \le A < \infty$, find

$$C^+(A) := \sup_{F\in\mathcal{A}^+} \frac{F(0) - A\int_{[-1,1]^c}(\widehat{F}(t))_+\,dt}{\|F\|_1}. \tag{4}$$

Here, we denote the L^1 norm $\|F\|_1 := \int_{\mathbb{R}} |F(x)|\,dx$, and $\widehat{F}(t)_+ = \max\{\widehat{F}(t), 0\}$. They show that, assuming RH, we may take the constant $C = C^+(36/11)^{-1}$ in the bound (3), giving this connection between a problem in Fourier analysis and the problem of bounding prime gaps. The specific parameter $A = \frac{36}{11}$ arises from other number-theoretic considerations. This Fourier optimization problem can be studied on its own, and they show several qualitative and quantitative properties. For example, they show that $1 \leq C^+(A) \leq 2$, for all $A \geq 1$, and furthermore, for any $A > 1$, there exists an extremal function F that yields the maximum value in (4).

While this problem can be stated in accessible terms, actually constructing an extremal function F, or even finding the exact value of $C^+(A)$ for a given parameter A, is a difficult problem. To make the constant in (3) as small as possible, the question is then to give good lower bounds for $C^+(36/11)$, by means of constructing a good explicit test function F. They find a specific example that gives the bound $C^+(A) > \frac{25}{21}$, improving the then best-known admissible value of C in (3) to $p_{n+1} - p_n \leq \left(\frac{21}{25} + o(1)\right)\sqrt{p_n}\log p_n$, and this is still the best approach to date to attack this problem. The numerics where slightly refined by Chirre et al. [10] by using strong computational techniques to construct test functions, via semidefinite programming. They replaced $\frac{21}{25}$ by 0.8358, and also used the method of Carneiro, Milinovich and Soundararajan to study gaps between primes in an arithmetic progression. Returning to [6], they also showed the following explicit version, with a similar but slightly larger constant: for all $x > 2$, there is always a prime number between x and $x + \frac{22}{25}\sqrt{x}\log x$.

In the article [11], we combine tools from Fourier analysis, analytic number theory and algebraic number theory to obtain several new estimates regarding integers and primes represented by quadratic forms. We have two main themes that are ubiquitous in this investigation. First, we use the well-known theme that propositions about quadratic forms can be stated in two other equivalent languages: ideals of number fields and lattices. We use all three points of view to our advantage in different parts of the article. Our second theme is the use of Fourier analysis, in the following way: we begin by finding a summation formula that connects our object of study with an arbitrary function and its Fourier transform; then, we choose an appropriate test function that recovers the desired information in an optimized manner. The use of a version of the Guinand-Weil explicit formula is one example of this (which is the focus of the present expository article), but not the only one. One of our results is an extension of the method of Carneiro, Milinovich, and Soundararajan to study gaps between primes represented by a given quadratic form. Consider, for instance, the quadratic form $q(x, y) = x^2 + 27y^2$, and let $p_{n,q}$ be the n-th prime number of the form $p = x^2 + 27y^2$. The following is a corollary of our results:

Theorem 3 *Assume GRH. Then,*

$$p_{n+1,q} - p_{n,q} \leq \left(\frac{6}{C^+(28)} + o(1)\right)\sqrt{p_{n,q}}\log p_{n,q}.$$

Moreover, we have the lower bound $C^+(28) \geq 1.0889$.

This immediately implies Theorem 1. To achieve this, we work in the language of ideals of imaginary quadratic fields, where we have the machinery of Hecke L-functions—another generalization of Dirichlet L-functions to this language—for which we can use an explicit formula similar to Lemma 1. To obtain the explicit bound for $C^+(28)$, we use the following test function in the optimization problem (4):

$$F(x) = H\left(\frac{x}{0.98644}\right),$$

where

$$H(x) = \cos(2\pi x)\left(\frac{68}{1-16x^2} + \frac{5}{9-16x^2} + \frac{1}{25-16x^2}\right).$$

In Sect. 6, we explain how to find such test functions numerically.

5 Least Quadratic Non-residue

Let us now turn our attention to the problem of bounding n_p, the least quadratic non-residue modulo a prime number p. For example, $1 = 1^2$ is always a quadratic residue. How about 2? Who is the smallest non-residue? Ankeny showed in 1952 that, assuming GRH, we have the bound $n_p \leq (C + o(1)) \log^2 p$, so that we can always find a non-residue that is much smaller than p. For further background, see [7] and the references therein. In [7], we consider the following optimization problem:

Extremal Problem 2
Find

$$C(1) := \sup_{0 \neq F \in \mathcal{A}} \frac{2\pi}{\|F\|_1} \left(\int_{-\infty}^{0} \widehat{F}(t) e^{\pi t}\, dt - \int_0^\infty |\widehat{F}(t)| e^{\pi t}\, dt \right), \qquad (5)$$

where the supremum is taken over the class of functions $\mathcal{A} = \{F : \mathbb{R} \to \mathbb{C};\ F \in L^1(\mathbb{R}) : \widehat{F} \text{ is real-valued }\}$.

We show the following connection:

Theorem 4 *Assume GRH. Let n_p be the least quadratic non-residue modulo a prime p. Then*

$$n_p \leq (C(1)^{-2} + o(1)) \log^2 p.$$

Furthermore, we have the bounds $1.146 < C(1) < 1.148$.

This immediately implies Theorem 2, and gives an improvement on the previous best asymptotic upper bounds under GRH, given by Lamzouri et al. [16]. An interesting feature of this result is that, since we have both lower bounds (obtained by constructing good functions f) and upper bounds (which are more subtle and require finer theoretical analysis of the optimization problem), we can see that we have essentially arrived at the limit of this method.

6 Numerically Optimizing the Bounds

Once we arrive at a Fourier optimization problem, it can be difficult to find an exact solution. The next step is to find good bounds to approximate it- that is, find a test function F that does a good job in the search for the supremum. A way to approach this is to start with a good family of test functions that parametrize our search space, and then use numerical optimization methods to find a good candidate within that family. One possible family is functions of the form $P(x)e^{-\pi x^2}$, where P is a polynomial: this family was used in [10], where candidates where found via semidefinite programming (see [1] for background on semidefinite programming). In other cases, bandlimited functions- those whose Fourier transform have compact support- are useful, such as in [11].

To illustrate this process, let us consider the extremal problem (5). To maximize the numerator, we should try to concentrate most of the mass of $\widehat{F}(t)$ on the negative numbers. With this in mind, we consider the Fourier transform pairs

$$F_n(x) = \frac{(2n-1)!}{\pi(1+2ix)^{2n}}; \qquad \widehat{F_n}(t) = -(\pi t)^{2n-1}\, e^{\pi t}\, \mathbf{1}_{\mathbb{R}_-}(t), \tag{6}$$

and note that $F_n(x) \in \mathcal{A}$ for all positive integers n. Here, $\mathbf{1}_{\mathbb{R}_-}(t)$ is the indicator function of the negative numbers, defined as 1 if $t < 0$ and 0 otherwise. We now consider the following family, composed by dilations and translations of linear combinations of functions as in (6), defined by:

$$\widehat{F}(t) = g\left(\frac{\pi t - c}{a}\right), \text{ where } g(t) = \sum_{n=1}^{N} b_n\, t^{2n-1}\, e^t\, \mathbf{1}_{\mathbb{R}_-}(t). \tag{7}$$

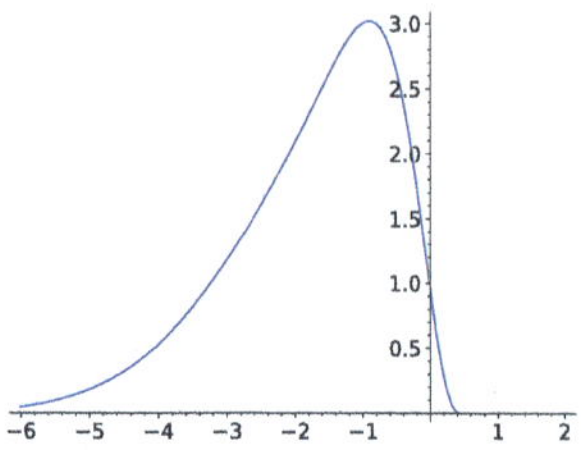

Fig. 1 Plot of $\widehat{F}(t)/\widehat{F}(0)$ defined in (8), a good test function for the optimization problem (5)

Here, $b_n, c \in \mathbb{R}$ and $a > 0$. Now we use numerical optimization methods to find good parameters. With the principal axis method of Brent [2], we find the function

$$\widehat{F}(t) = g\left(\frac{t - 0.47}{0.42}\right), \tag{8}$$

where

$$g(x) = 2e^x \left(0.0006x^7 + 0.0005x^5 + x^3 + 0.0405x\right) \mathbf{1}_{\mathbb{R}_-}(x).$$

See Fig. 1 for a plot of $\widehat{F}(t)$. This yields the bound $1.143 < C(1)$, which is close to what we claim. In [7], we use similar, more complicated examples to get our result.

Acknowledgments The author was supported by a PIMS-Simons postdoctoral fellowship at the University of Lethbridge.

References

1. Blekherman, G., Parrilo, P.A., Thomas, R.R. (eds.) Semidefinite Optimization and Convex Algebraic Geometry, volume 13 of MOS-SIAM Series on Optimization (Society for Industrial and Applied Mathematics (SIAM)/Mathematical Optimization Society, Philadelphia, 2013)
2. Brent, R.P.: Algorithms for Minimization Without Derivatives. Prentice-Hall Series in Automatic Computation. Prentice-Hall, Englewood Cliffs (1973)
3. Carneiro, E., Chirre, A.: Bounding $S_n(t)$ on the Riemann hypothesis. Math. Proc. Camb. Philos. Soc. **164**(2), 259–283 (2018)
4. Carneiro, E., Chandee, V., Milinovich, M.B.: Bounding $S(t)$ and $S_1(t)$ on the Riemann hypothesis. Math. Ann. **356**(3), 939–968 (2013)
5. Carneiro, E., Chirre, A., Milinovich, M.B.: Bandlimited approximations and estimates for the Riemann zeta-function. Publ. Mat. Barc. **63**(2), 601–661 (2019)
6. Carneiro, E., Milinovich, M.B., Soundararajan, K.: Fourier optimization and prime gaps. Comment. Math. Helv. **94**(3), 533–568 (2019)
7. Carneiro, E., Milinovich, M.B., Quesada-Herrera, E., Ramos, A.P.: Fourier optimization, the least quadratic non-residue, and the least prime in an arithmetic progression. Preprint. Available at https://arxiv.org/abs/2404.08380
8. Carneiro, E., Milinovich, M.B., Ramos, A.P.: Fourier optimization and Montgomery's pair correlation conjecture. Math. Comput. **94**, 409 (2024)

9. Chandee, V., Soundararajan, K.: Bounding $|\zeta(\frac{1}{2}+it)|$ on the Riemann hypothesis. Bull. Lond. Math. Soc. **43**(2), 243–250 (2011)
10. Chirre, A., Pereira Júnior, V.J., de Laat, D.: Primes in arithmetic progressions and semidefinite programming. Math. Comput. **90**(331), 2235–2246 (2021)
11. Chirre, A., Quesada-Herrera, E.: Fourier optimization and quadratic forms. Q. J. Math. **73**(2), 539–577 (2022)
12. Cohn, H., Elkies, N.: New upper bounds on sphere packings. I. Ann. Math. (2) **157**(2), 689–714 (2003)
13. Cohn, H., Kumar, A., Miller, S.D., Radchenko, D., Viazovska, M.: The sphere packing problem in dimension 24. Ann. Math. (2) **185**(3), 1017–1033 (2017)
14. Davenport, H.: Multiplicative Number Theory, vol. 74, 2nd edn. Springer, New York (1980)
15. Iwaniec, H., Kowalski, E.: Analytic Number Theory, volume 53 of American Mathematical Society Colloquium Publications. American Mathematical Society, Providence (2004)
16. Lamzouri, Y., Li, X., Soundararajan, K.: Conditional bounds for the least quadratic non-residue and related problems. Math. Comput. **84**(295), 2391–2412 (2015)
17. Montgomery, H.L., Vaughan, R.C.: Multiplicative Number Theory: I. Classical Theory. Cambridge Studies in Advanced Mathematics, vol. 97. Cambridge University Press, Cambridge (2006)
18. Quesada-Herrera, E.: On the q-analogue of the pair correlation conjecture via Fourier optimization. Math. Comput. **91**(337), 2347–2365 (2022)
19. Riemann, B.: Ueber die anzahl der primzahlen unter einer gegebenen grosse. Ges. Math. Werke Wissensch. Nachlaß **2**(145–155), 2 (1859)
20. Titchmarsh, E.C.: The Theory of the Riemann Zeta-function, 2nd edn., rev. by D. R. Heath-Brown, vol. x, 412 pp. Oxford Science Publications/Clarendon Press, Oxford (1986)
21. Vaaler, J.D.: Some extremal problems in Fourier analysis. Bull. Amer. Math. Soc. **12**(2), 183–216 (1985)
22. Viazovska, M.S.: The sphere packing problem in dimension 8. Ann. Math. (2) **185**(3), 991–1015 (2017)

Quantile Regression in the Analysis of the Distribution of Total Monthly Income

Lesky Ibeth Rivas Martínez

Abstract Quantile regression is a statistical technique that models the relationship between variables for different quantiles of the conditional distribution of the dependent variable, not just the mean. Unlike traditional linear regression, which estimates the conditional mean, quantile regression estimates parameters for specific quantiles. This provides a more detailed view of how the relationship between variables changes at different points in the distribution, which is useful in the presence of outliers or skewed distributions. The technique minimizes a weighted loss function based on the desired quantile, thus offering better robustness. In this paper, a quantile regression model is used to explore the relationship between sex, age, and educational level at different points in the distribution of total monthly income. Using the permanent household survey database of the National Institute of Statistics of Honduras (INE), the model results demonstrate that the effects of these variables vary significantly between different quantiles of the income distribution.

Keywords Quantile regression · Income distribution · Conditional distribution

1 Introduction

In the classical ordinary least squares (OLS) methodology, the conditional mean function is fundamental. This function describes how the mean of the response variable (y) varies with respect to the covariates (x). Understanding this relationship is crucial for obtaining the best unbiased linear estimates through ordinary least squares, which requires certain assumptions to be met: linearity, independence of errors, and homoscedasticity. These assumptions provide the basis for interpreting the results and conclusions of the model.

L. I. Rivas Martínez (✉)
Universidad Nacional Autónoma de Honduras, Tegucigalpa, Honduras

P. E. Harris et al. (eds.), *Women in Mathematics in Latin America*,
Research Perspectives Ghent Analysis and PDE Center 14,
https://doi.org/10.1007/978-3-031-97857-9_16

In many cases, the assumptions of the OLS method are not met. This is where quantile regression comes in, which posits the relationship between the response variable and the independent variables at any quantile of the conditional distribution function. Thus, quantile regression can provide complete information about the relationship between the response variable (y) and the covariates (x) across the entire conditional distribution,[1] without making any distributional assumptions about the error term in the model [1].

Quantile regressions are very useful for visualizing changes in the conditional distribution of longitudinal data sets over time (e.g., heteroscedasticity[2] in the data set). When the mean regressions are indeed significantly different, quantile regression can suggest which part of the conditional distributions differ.

Quantile regression is used in many fields, such as in pediatrics, which allows, for example, the determination of growth curves for children's height and weight. It is also widely used in economics to study wages, discriminatory effects, and income inequality trends.

The objective of this article is to determine whether gender, age, and educational level influence different points in the distribution of monthly income using a quantile regression model, to identify and analyze the variations in the impact of these variables across different income percentiles

2 Quantile Regression

The objectives pursued in quantile regression are similar to those of linear regression by Ordinary Least Squares (OLS), that is, to model the relationship between variables. However, due to factors such as heteroscedasticity, the presence of outliers, and structural changes, the mean value estimated by OLS for the endogenous variable is not always the most representative. Thus, quantile regression offers the possibility of creating different regression lines for different quantiles of the endogenous variable through an estimation method that is less affected by the aforementioned reasons.

2.1 Preliminaries

Just as OLS regression is linked to the mean, quantile regression, as its name indicates, is based on the concept of quantile. The classical quantile regression model was introduced in 1978 by Koenker and Bassett [2] as an extension of the notion of ordinary quantiles (also called "percentiles") in a location model,

[1] Conditional distribution: the distribution of a subset of data given a specific condition.

[2] Heteroscedasticity: the variance of the errors of a model is not the same for all observations.

expanding it to a more general class of linear models where the conditional quantiles take a linear form [3].

To recall the concept of ordinary quantile, suppose we have a sample of observations of a variable Y with a distribution $F(.)$, $Y_t : t = 1, 2, \ldots, N$.

The quantile τ of the sample, with $0 < \tau < 1$, is that value b which leaves a proportion τ of observations below b and a proportion $(1 - \tau)$ above. In the case of the median $\tau = 0.5$, 50% of the data will be below b and 50% of the data will be above [4].

An alternative way to express the definition of quantiles, which is also a first approximation to the estimation method of quantile regression, is given by the following expression:

$$\min_{b \in \mathbb{R}} \left[\sum_{y_i \geq b} \tau |y_i - b| + \sum_{y_i < b} (1 - \tau)|y_i - b| \right]. \quad (1)$$

Where τ is the quantile, y_i are the sample observations for the variable y and b are the values that minimize the expression [4].

2.2 *Quantile Regression Model*

Quantile regression extends the linear regression model to estimate the conditional quantiles of the response variable. Unlike traditional linear regression, which estimates the conditional mean, quantile regression focuses on the $\tau - quantile$ of the dependent variable y_i given the covariates x_i. This approach provides a more comprehensive understanding of the relationship between the variables, particularly in the presence of heteroscedasticity or outliers.

According to Koenker and Bassett [2], the quantile regression model is expressed by an equation similar to the linear regression model given by:

$$y_i = \beta_0 + \beta_1 x_{i1} + \ldots + \beta_p x_{ip} + \epsilon_i. \quad (2)$$

Where y_i is the endogenous or dependent variable, x_i are the exogenous or explanatory variables β are the fixed and unknown parameters and ϵ_i is the estimation error.

For the linear regression model, the conditional mean of y_i given $x_{i1} \ldots, x_{ip}$ is $E(y_i|x_{i1}, \ldots, x_{ip}) = \beta_0 + \beta_1 x_{i1} + \ldots + \beta_p x_{ip}$ which implies that the error term ϵ_i has an expectation of zero. The quantile regression model can be equivalently written as:

$$Q_\tau(y_i|x_{i1}, \ldots, x_{ip}) = \beta_{0,\tau} + \beta_{1,\tau} x_{i1} + \ldots + \beta_{p,\tau} x_{ip} + Q_\tau(\epsilon_i). \quad (3)$$

This implies that the $\tau - quantile$ of the error term ϵ_i is zero, meaning the errors are distributed in such a way that their $\tau - quantile$ is zero.

In the quantile regression model, a relationship is established between the $\tau - quantile$ and the covariates. Thus, the conditional quantile given $x_{i1}, \ldots, x_{ip}$ is:

$$Q_y(\tau|x_{i1}, \ldots, x_{ip}) = \beta_{0,\tau} + \beta_{1,\tau}x_{i1} + \ldots + \beta_{p,\tau}x_{ip}. \tag{4}$$

2.2.1 Estimation

OLS regression is limited to estimating a specific location of the conditional distribution of y, namely the conditional mean. In contrast, quantile regression expands this approach by allowing the analysis of the conditional distribution of y at various points, corresponding to different quantiles. Thus, the quantile regression estimates the conditional quantile of y given x by solving the following minimization problem:

$$\min_{u \in \mathbb{R}} \left[\sum_{i=1}^{n} \rho_\tau(y_i - u) \right]. \tag{5}$$

The check function ρ_τ is defined as:

$$\rho_\tau(u) = \begin{cases} \tau u \text{ if } u \geq 0 \\ (\tau - 1)u \text{ if } u < 0. \end{cases} \tag{6}$$

This can be graphically represented as shown in Fig. 1, where the slope of the line relating $\rho_\tau(u)$ and u is τ when the residuals are positive and $\tau - 1$ otherwise.

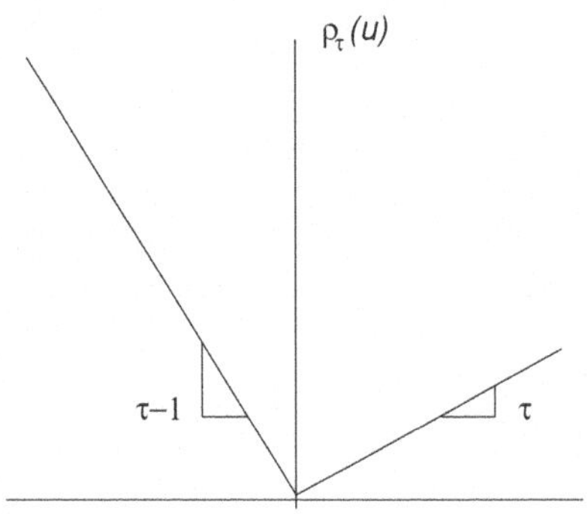

Fig. 1 Quantile regression function ρ. Source: Koenker and Hallock [5]

The conditional quantile function τ as $Q_y(\tau|x) = x^T \beta(\tau)$, and the solution of $\hat{\beta}(\tau)$ can be considered [6]:

$$\min_{\beta \in \mathbb{R}^p} \left[\sum_{i=1}^{n} \rho_\tau (y_i - x_i^T \beta) \right]. \tag{7}$$

The resulting minimization problem in (7) can be reformulated as a linear programming problem as follows:

$$\min_{(\beta, u, v) \in \mathbb{R}^p \times \mathbb{R}_+^{2n}} \left[\tau \mathbf{1}_n^T u + (1 - \tau) \mathbf{1}_n^T v \,\middle|\, X\beta + u - v = y \right]. \tag{8}$$

Where X now denotes the $n \times p$ matrix. The residual vector $y - X\beta$ has been divided into its positive and negative parts, thus minimizing a linear function. Most of the important properties of the solutions, $\hat{\beta}(\tau)$, called quantile regressions, are immediately derived from known properties of linear programming solutions.

In general, it is important to consider the using of quantile regression when the assumptions of linear regression are violated, as quantile regression can offer a more robust solution.

3 Application

The database used in this study comes from the June 2023 Permanent Household Survey of Multiple Purposes (EPHPM) of the National Institute of Statistics of Honduras (INE) [7].

For the application of the quantile regression model, only the economically active employed population was selected from the database, with the variables age, gender, and educational level. Table 1 presents the description of the variables.

Model A quantile regression model was proposed to evaluate the effect of the factors that influence the distribution of the total monthly income of the employed population in Honduras.

$$log(y_i) = \beta_{0,\tau} + \beta_{1,\tau} X1 + \beta_{2,\tau} X2 + \beta_{3,\tau} X3 + \epsilon_{i,\tau} \tag{9}$$

Table 1 Description of variables

Variable	Description
log(y)	Logarithm of total monthly income
X1	Working age population, people over 15 years old
X2	Sex: 0 = female, 1 = male
X3	Educational level: 1 = no level, 2 = primary, 3 = secondary, 4 = higher

Source: Prepared by the author on the basis of data supplied by INE [7]

Table 2 Estimation results for $\tau = 0.2$

Variables	Estimation	Standard error	t value	pr(>|t|)
(Intercept)	4.94103	0.21756	22.71073	0.00000
X1	0.00725	0.00299	2.42670	0.01533
X2	1.04133	0.08532	12.20528	0.00000
X3	0.68414	0.05820	11.75438	0.00000

Source: Own elaboration

Table 3 Estimation results for $\tau = 0.9$

Variables	Estimation	Standard error	t value	pr(>|t|)
(Intercept)	8.18669	0.19667	41.62667	0.00000
X1	0.00527	0.00264	1.99294	0.04641
X2	0.71473	0.07802	9.16038	0.00000
X3	0.38777	0.04984	7.77983	0.00000

Source: Own elaboration

Where y_i is the total monthly income, X are the covariates or independent variables, β_τ are the parameters specific to the τ-quantile, and finally $\epsilon_{i,\tau}$ represents the error corresponding to that quantile. To establish the correct transformation of total monthly income, the logarithm function was used, allowing for better interpretation in the data analysis.

With the results of the estimates shown in Tables 2 and 3, the lower end at the 0.2 quantile and the upper end at the 0.9 quantile were analyzed.

For the age variable ($X1$), there is a positive, albeit small, effect. Each additional year of age slightly increases monthly income, but the impact is less pronounced compared to the 0.9 quantile.

The magnitude of the gender effect is smaller at the 0.9 quantile compared to the 0.2 quantile. This indicates that the gender income gap is smaller at the upper end of the income distribution compared to the lower end.

The educational level at the 0.2 quantile has a stronger effect on income. An increase in the educational level is associated with a greater improvement in monthly income compared to the 0.9 quantile, where the return on educational level in terms of income is lower.

In Fig. 2, each point represents the estimated regression coefficient for a quantile, the solid red line is the estimated regression coefficient for the predictor using ordinary least squares, and the dashed lines are the 95% confidence limits. These results show how the explanatory variables $X1$, $X2$, and $X3$ affect different parts of the conditional distribution of y. The magnitudes and significances vary between the quantiles, offering a more detailed and robust view of the impact of these variables on y compared to OLS.

Figure 2 also shows that the impact of age ($X1$) on monthly income becomes significant starting from the 0.3 quantile. This indicates that the effect of age on

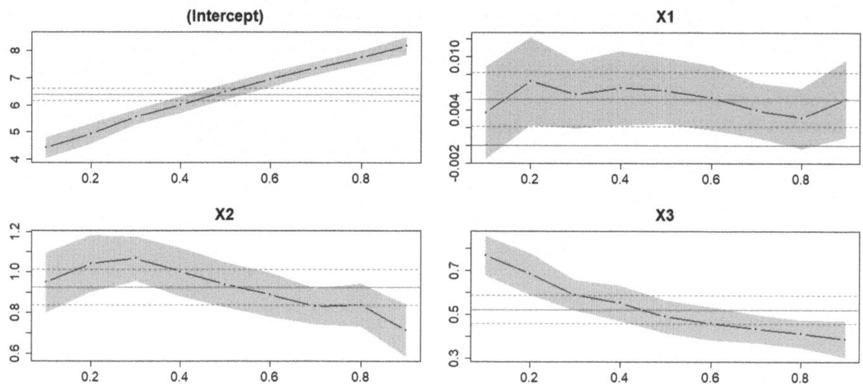

Fig. 2 Quantile regression coefficients Source: Own elaboration

income is small in the lower quantiles where incomes are low, but it becomes more relevant at higher income levels.

For the gender variable ($X2$), the coefficient is positive and significant across all quantiles, although it slightly decreases as the quantile increases. This indicates that, on average, men have a higher monthly income than women. This effect is consistent across all quantiles but is more pronounced at the lower quantiles.

The coefficient for the educational level ($X3$) is significant across all quantiles and decreases as the quantile increases. This indicates that the effect of educational level on income is more pronounced in the quantiles where monthly incomes are low and tends to decrease in the quantiles where monthly incomes are high.

3.1 Concluding Remarks

Unlike ordinary least squares linear regression, which focuses on the mean, quantile regression allows us to examine how the independent variables: age, gender, and educational level affect different points in the income distribution.

This technique provides a more detailed understanding by revealing how the impact of these variables varies according to the income percentile, potentially showing a more significant effect at higher percentiles. Quantile regression is a powerful tool for analyzing how variables influence monthly income throughout the entire distribution.

References

1. John, O., Nduka, E.: Quantile regression analysis as a robust alternative to ordinary least squares. Sci. Afr. **8**(2), 61–65 (2009)

2. Koenker, R., Bassett Jr., G.: Regression quantiles. Econometrica **46**, 33–50 (1978)
3. Buhai, S.: Quantile regression: overview and selected applications. Ad Astra **4**(4), 1–17 (2005)
4. Otero, J.V., Reyes, B.S.: Regresión Cuantílica: estimación y contrastes. Instituto LR Klein-Centro Gauss (2012)
5. Koenker, R., Hallock, K.F.: Quantile regression. J. Econ. Perspect. **15**(4), 143–156 (2001)
6. Koenker, R.: Quantile Regression. Econometric Society Monographs, No. 38. Published in the United States of America by Cambridge University Press, New York (2005)
7. Instituto Nacional de Estadística Honduras (INE). https://ine.gob.hn/v4/

The k-Yamabe Flow and Its Solitons

Maria Fernanda Espinal and Mariel Sáez-Trumper

Abstract This chapter announces our upcoming result on k-Yamabe solitons. These are special solutions to a fully non-linear geometric flow that extends the well-known Yamabe flow. The equation that we are proposing has only been understood in some special cases (specific choices of parameters and compact locally conformally flat manifolds) and the study of solitons is relevant to extend this understanding, particularly in describing possible singularities that may form during the evolution. Our contribution is a first step in that direction.

Keywords k-Yamabe solitons · Singularities · Conformal geometry

1 The Yamabe Problem and the Yamabe Flow

The Yamabe problem is a classical question in conformal geometry that seeks for existence of metrics with constant scalar curvature within a conformal class. The problem was posed by H. Yamabe in 1960 [23] as a possible extension of the famous uniformization theorem, which states that every simply connected Riemann surface is conformally equivalent to the open unit disk, the complex plane, or the Riemann sphere. This statement can be translated into a curvature condition that asserts that the universal cover of every Riemann surface can be endowed with a metric of constant Gaussian curvature.

M. F. Espinal (✉)
CMM-Center for Mathematical Modeling, Santiago, Chile
e-mail: mfespinal@cmm.uchile.cl

M. Sáez-Trumper
P. Universidad Católica de Chile, Facultad de Matemáticas, Santiago, Chile
e-mail: mariel@uc.cl

P. E. Harris et al. (eds.), *Women in Mathematics in Latin America*,
Research Perspectives Ghent Analysis and PDE Center 14,
https://doi.org/10.1007/978-3-031-97857-9_17

To rigorously state the Yamabe problem, we consider (M^n, g), a closed Riemannian manifold of dimension $n \geq 3$. The family of metrics that are conformal to a given background g are the metrics of the form

$$g_u = u^{\frac{4}{n-2}} g, \text{ with } u > 0.$$

The question posed by Yamabe is whether is possible to find a metric g_u with constant scalar curvature. This is equivalent to finding a function u that solves the equation

$$-\frac{4(n-1)}{n-2} \Delta_g u + \mathrm{R}_g u = K u^{\frac{n+2}{n-2}} \quad \text{in} \quad M, \tag{1}$$

where Δ_g is the Laplace-Beltrami operator associated with the Riemannian metric g, R_g is the scalar curvature with respect to this metric and K is the constant that prescribes the curvature.

Although Eq. (1) has a variational structure, the exponent $\frac{n+2}{n-2}$ is critical with respect to Sobolev embeddings and this imposes several challenges in its study. Nonetheless, a positive answer to the conjecture was obtained by series of works of Yamabe [23], Trudinger [21], Aubin [1] and Schoen [19].

After the conjecture had been already confirmed, an alternative approach was proposed by R. Hamilton in [14]. He suggested to use a geometric flow, which is now known as the *Yamabe flow*, or equivalently, to study solutions to the parabolic equation

$$\begin{cases} \frac{d}{dt} g & = -\mathrm{R}_g g \\ g(0) & = g_0, \end{cases} \tag{2}$$

with

$$g(\cdot, t) = u^{\frac{4}{n-2}}(\cdot, t)\, g_0(\cdot) \quad \text{and} \quad u > 0. \tag{3}$$

In general, parabolic equations are expected to have a regularizing effect and it is presumed that their solutions will converge to static ones as $t \to \infty$. In the particular case of the Yamabe flow, this would mean that limiting solutions are Yamabe metrics (of constant scalar curvature) that are conformally equivalent to the initial condition. The conjecture of Hamilton was confirmed in the locally conformally flat case by works of B. Chow [7] and R. Ye in [24]; the more general setting was later completed by S. Brendle in [2, 3].

For non-compact manifolds there are counterexamples to the Yamabe conjecture [15]; however it still holds if additional conditions are imposed (see for instance [20]). From the viewpoint of the flow, this translates into solutions that develop a finite time singularity. An example of the latter phenomenon was presented in [9], where the authors also show that a one parameter family obtained by suitable parabolic rescalings at the singular time converges to a special solution know as the Barenblatt solution. It is still an open problem to understand for which (non-compact) initial conditions solutions develop singular behavior and what are the possible models for these singularities.

The Barenblatt solution belongs to a larger class of metrics known as *solitons*. More precisely, a metric $g(t)$ is called a *soliton solution* if there exists a smooth function $\alpha(t)$ and 1-parameter family of diffeomorphisms $\{\phi_t\}$ of M such that

$$g(t) = \alpha(t)\phi_t^*(g_0), \tag{4}$$

with $\alpha(0) = 1$ and $\phi_0 = id_M$. These solitons are known as *gradient* if the vector field X generated by the 1-parameter family ϕ_t arises as the gradient of a function φ. Moreover, a soliton is called shrinking, steady or expanding if $\rho := -\alpha'(0)$ is positive, zero or negative, respectively.

A general philosophy in the study of singularity formation for geometric flows is based on the observation that a suitable parabolic rescaling of a solution produces a new solution to the equation. When these rescalings are performed around a singular point in space-time, appropriate choices of parameters and compactness properties imply the convergence of the family of rescalings to a soliton, which may be shrinking, steady or expanding, depending on the *speed* at which the singularity forms. This approach has been successfully performed for many flows, for instance the Ricci Flow (see Chapter 8 in [8]) and the Mean Curvature Flow (see Chapters 3 and 4 in [18]); however, in the case of the Yamabe Flow the understanding of this procedure is not as developed as in the previous examples. Finally, to complete the analysis of singularities, it is necessary to classify the possible soliton solutions that emerge after rescaling. For the Yamabe Flow this was achieved by P. Daskalopoulos and N. Sesum in [10], under the additional assumption that the solutions are conformally flat gradient solitons.

2 The k-Yamabe Problem

In dimensions higher than 2 the control of the scalar curvature seems to bear a weaker connection with the underlying topology; however, other curvature quantities may play a bigger role. To explain this statement, let (M^n, g) be a complete, connected smooth Riemannian manifold of dimension $n \geq 3$. Consider Ric_g and R_g to be the Ricci tensor and scalar curvature of g, respectively and define the Schouten tensor with respect to the metric g as

$$A_g = \frac{1}{n-2}\left(\mathrm{Ric}_g - \frac{\mathrm{R}_g}{2(n-1)}g\right). \tag{5}$$

For $n = 4$ the Chern-Gauss-Bonnet formula is equivalent to

$$8\pi^2\chi(M) = \int_M \left(\frac{1}{4}|W_g|^2 + \sigma_2(g)\right) dv_g, \tag{6}$$

where $\chi(M)$ is the Euler characteristic of M, W is the Weyl tensor and σ_2 is the *second symmetric polynomial* associated to the Schouten tensor. The first integral term in Eq. (6) is known to be invariant under conformal changes and $\chi(M)$ is a topological invariant, hence $\int_M \sigma_2(g)dv_g$ is also invariant under conformal changes. The precise definition of the quantity above is $\sigma_2(g) = \sum_{i<j} \lambda_i \lambda_j$, where $\{\lambda_i\}_{i=1}^n$ are the eigenvalues of the tensor $g^{-1}A_g$. Since the Ricci tensor is symmetric, these eigenvalues are n real numbers (not necessarily distinct) and $\sigma_2(g)$ is well defined.

In general, for $k \in \{1, \ldots, n\}$ we define the σ_k-curvature as the k-th elementary symmetric function on the eigenvalues of the tensor $g^{-1}A_g$:

$$\sigma_k(g) := \sigma_k(g^{-1}A_g) = \sum_{i_1<i_2<\ldots<i_k} \lambda_{i_1} \ldots \lambda_{i_k}, \quad \text{for} \quad 1 \le k \le n. \tag{7}$$

In particular, $\sigma_1(g)$ is the trace of $g^{-1}A_g$ and, in fact, the σ_1-curvature is equal to a multiple of the scalar curvature R_g.

Then, the *k-Yamabe problem* is to find metrics that satisfy

$$\sigma_k(g) = K, \tag{8}$$

for a given constant K. In particular, for $k = 1$ it agrees with the classical Yamabe problem. From (6) we can observe that for $n = 4$ and $k = 2$ possible solutions are connected to the underlying topology.

From the view point of partial differential equations, this is a fully-nonlinear problem which in general cannot be guaranteed to be elliptic, unless additional restrictions are imposed. A classical condition is to consider metrics within the *positive Garding cone* which is given by

$$\Gamma_k^+ = \{g : \sigma_1(g), \ldots, \sigma_k(g) > 0\}. \tag{9}$$

If $g \in \Gamma_k^+$ we say that g is *k-admissible.* There are other possible conditions that ensure ellipticity, such as considering metrics in the *negative cone*, but we will not address them in this work. It is worth remarking that although ellipticity is not an issue when $k = 1$, the difficulties in the study of the Yamabe problem are connected to the sign of the expected constant value for scalar curvature; for instance, when the constant is negative, it is possible to apply the maximum principle, which simplifies the analysis.

The k-Yamabe problem has been studied by many authors, often in analogy with the classical case ($k = 1$). Several results have been obtained under two assumptions on the background metric g: that is k-admissible and locally conformally flat. For instance, under these assumptions, J. Viaclovsky [22] showed that, as in the case of $k = 1$, solutions to (8) can be obtained from variational methods if $n \neq 2k$. A second example of results in this situation can be found in the work of A. Li and Y.Y. Li in [16], where the authors studied (8), obtaining Liouville-type theorems, existence and compactness in some situations; these results were also known for $k = 1$. Parabolic versions of the k-Yamabe problem were also address under the hypotheses of admissibility and locally conformally flatness, for instance, when $n \neq 2k$, P. Guan and G. Wang considered a related flow in [13], showing long-time existence and the sequential convergence of solutions as $t \to \infty$. The case $n = 2k$ was treated by S. Brendle and J. Viaclovsky in [4] using different methods. There are fewer results when the locally conformally flat assumption is removed, but there are some remarkable examples in the case $k = 2$: for $n \neq 4$, J. Viaclovsky [22] proved that solutions to (8) have a variational structure, while when $n = 4$ a priori estimates and existence was established by A. Chang et al. in [6]. The admissibility condition has also been replaced in some works, see for instance [11] and references therein for the elliptic case and [17] for a related parabolic equation.

Note that in that in the previous discussion often the cases $n \neq 2k$ and $n = 2k$ were considered separately. In fact, these cases can be separated further into $n > 2k$, $n = 2k$ and $n < 2k$. While classical Yamabe falls in the first category and many analogies can be drawn, in the other two cases some differences are observed. For instance, when M is locally conformally flat and $n \neq 2k$ solutions to (8) are critical points of the functional $\int \sigma_k(A(g))dv_g$ (see [22]), while this integral is a conformal invariant if $n = 2k$. Furthermore, when $n \leq 2k$ admissibility implies the positivity of the Ricci tensor, and therefore (M, g) is conformally covered by $\mathbb{S}^n$, and both existence and compactness follow from known results.

3 Main Results

In [12] we propose to study the following geometric flow

$$\begin{cases} \frac{d}{dt} g &= -\sigma_k^{1/k}(g) g \\ g(0) &= g_0, \end{cases} \tag{10}$$

with

$$g(\cdot, t) = u^{\frac{4k}{n+2k}}(\cdot, t) g_0(\cdot) \quad \text{and} \quad u > 0. \tag{11}$$

In our work we study gradient soliton solutions to Eq. (10) that are conformal to the flat metric. The definition of soliton solutions is given by (4) and the problem reduces to the following equation

$$(\sigma_k^{1/k}(g) - \rho)g = \frac{1}{2}\mathcal{L}_X g, \tag{12}$$

where $\rho = -\alpha'(0)$, X is the vector field generated by the 1-parameter family ϕ_t and $\mathcal{L}_X$ denotes the Lie derivative with respect to X. As before, if X is the gradient of a function φ then $g(t)$ is called a *gradient soliton* and in this setting, $\frac{1}{2}\mathcal{L}_X g = \nabla^2\varphi$. Here ∇ denotes the Levi-Civita connection with respect to the metric g.

In [10] the authors showed that, under suitable curvature assumptions (as positive sectional curvature), locally conformally flat Yamabe solitons are rotationally symmetric. This result was extended in [5], where the authors show that one can remove the assumption of local conformal flatness and relax the hypothesis on the sectional curvature. Specifically, the authors proved that any complete, noncompact, gradient k-Yamabe soliton (M^n, g) with nonnegative Ricci tensor is either a direct product $\mathbb{R} \times N^{n-1}$ where (N^{n-1}, g_N) is an $(n-1)$-dimensional complete Riemannian manifold with nonnegative Ricci tensor, or (M^n, g) is rotationally symmetric and globally conformally equivalent to $\mathbb{R}^n$ (Theorem *3.6* in [5]). With this result in mind we assume that our solutions are radially symmetric and globally conformally equivalent to $\mathbb{R}^n$.

From this perspective, the aim of our work is to extend the results in [10] to solutions of (12) that are gradient, conformally flat and rotationally symmetric. Our first result reduces the classification of conformally flat rotationally symmetric k-Yamabe solitons to the classification of global smooth solutions of a fully nonlinear elliptic equation.

Theorem 1 (PDE Formulation of Yamabe Solitons) *Let g_u defined by*

$$g_u(\cdot, t) = u^{\frac{4k}{n+2k}}(\cdot, t)|dx|^2, \quad u > 0. \tag{13}$$

Then g_u is a conformally flat rotationally symmetric k-Yamabe gradient soliton with $\sigma_k(g_u) > 0$ if and only if u is a smooth radial solution to the elliptic equation

$$\begin{aligned}\sigma_k^{1/k}&\left(D^2(u^{-\frac{2k}{n+2k}}) - \frac{u^{-\frac{2k}{n+2k}}}{2}|D(u^{-\frac{2k}{n+2k}})|^2 I\right)\\ &= u^{\frac{2k}{n+2k}-1}\left[(2\theta+\rho)u + \frac{4k\theta}{n+2k}x \cdot Du\right],\end{aligned} \tag{14}$$

where θ is a parameter that satisfies $2\theta + \rho > 0$. Equation (14) equivalent to

$$D_j(D_i u^{-\frac{2k}{n+2k}} T_{k-1}^{ij}) - n u^{\frac{2k}{n+2k}} T_{k-1}^{ij} D_i u^{-\frac{2k}{n+2k}} D_j u^{-\frac{2k}{n+2k}}$$
$$+ u^{\frac{2k}{n+2k}} \frac{n-k+1}{2} \sigma_{k-1} |D u^{-\frac{2k}{n+2k}}|^2$$
$$= k u^{\frac{k(n-2k)}{n+2k}} \left[(2\theta + \rho) u + \frac{4k\theta}{n+2k} x \cdot D u \right]^k ,$$

where T_{k-1} is the $(k-1)$-th Newton tensor evaluated on the Schouten tensor $g_u^{-1} A_{g_u}$. Here D denotes the derivative with respect to the Euclidean metric.

Expression (14) allow us to provide the following classification result for smooth, admissible, conformally flat, radially symmetric solitons. As expected, existence depends on the relation between n and $2k$ and our first result considers the case of $n \geq 2k$ and a suitable relation between the parameter θ (related to the gradient nature of our soliton) and ρ (that determines whether the solution is shrinking, steady or expanding).

Theorem 2 (Existence of Radial k-Yamabe Solitons for $n \geq 2k$) *Let $n \geq 2k$. For every $\alpha > 0$ the elliptic equation (14) admits a non-trivial admissible radially symmetric smooth solution u_α that satisfies $u_\alpha(0) = \alpha$ if and only if*

$$\theta \geq 0 \quad \text{and} \quad 2\theta + \rho > 0.$$

We remark that a solution u is admissible if the associated metric g_u belongs to the positive cone Γ_k^+.

As previously explained, the sign of ρ determines whether the solitons are shrinking, steady or expanding, which in terms of the blow-up procedure, is related to the speed at which the singularity forms. This translates into different asymptotic behaviors at infinity, which depend on the values of ρ, θ, n and k. These behaviors are summarized by the following theorem.

Theorem 3 (Asymptotic Behavior) *Let $m = \frac{n-2k}{n+2k}$, $n \geq 2k$ and ρ, θ as in Theorems 1 and 2. Let $\{u_\alpha\}$ be the family of solutions given by Theorem 2. Then the following holds.*

1. ***Yamabe expander** $\rho < 0$: It holds $u_\alpha(x) = O(|x|^{-2-\delta})$ as $|x| \to \infty$, where $\delta = \frac{\rho}{2k\theta(1-m)}$.*
2. ***Yamabe steady** $\rho = 0$:*
   - *For $n > 2k$ the decay rate at infinity is given by $u_\alpha(x) = O\left(\left[\frac{\ln|x|}{|x|^2}\right]^{\frac{1}{1-m}}\right)$ as $|x| \to \infty$.*
   - *If $n = 2k$ we have $u_\alpha(x) = O\left(\frac{(\ln|x|)^{1-\frac{2}{n}}}{|x|^2}\right)$ as $|x| \to \infty$.*

3. ***Yamabe shrinker*** $\rho > 0$*:*

- *If* $n > 2k$ *and* $0 < \rho \leq 2\theta$ *then solutions have a slow-decay rate at infinity, namely* $u_\alpha(x) = O(|x|^{-\frac{2}{1-m}})$ *as* $|x| \to \infty$.
- *If* $n > 2k$ *and* $0 < 2\theta < \rho$ *solutions either have a slow-decay rate at infinity* $u_\alpha(x) = O(|x|^{-\frac{2}{1-m}})$ *as* $|x| \to \infty$ *or* $u_\alpha(x) = O(|x|^{-\frac{4}{1-m}})$ *as* $|x| \to \infty$.
- *If* $n = 2k$ *then* $u_\alpha(x) = O(|x|^{-2(1+d)})$ *as* $|x| \to \infty$ *for some* $0 < d \leq \min\{\frac{\rho}{2\theta}, 1\}$.

We remark that for $\rho > 2\theta$ we do not know in which cases each behavior occurs, nor if in the last statement all values of d are possible.

Finally, when $n < 2k$ existence is not expected in every case. More precisely, we have the following result.

Theorem 4 (Non-existence of Radial k-Yamabe Solitons for $n < 2k$) *Let* $n < 2k$ *and* $m = \frac{n-2k}{n+2k} < 0$, *then the elliptic equation* (14) *does not have an admissible solution for* $\rho < 2\theta$.

For $\rho \geq 2\theta$ *and* $\alpha > 0$, *there exists a one parameter family* u_α *of smooth radially symmetric admissible solutions of Eq.* (14) *on* $\mathbb{R}^n$ *and their decay rate is given by* $u_\alpha(x) = O(|x|^{-\frac{4}{1-m}})$ *as* $|x| \to \infty$.

References

1. Aubin, T.: Equations differentielles non lineaires et probleme de Yamabe concernant la courbure scalaire. J. Math. Pures Appl. **55**, 269–296 (1975)
2. Brendle, S.: Convergence of the Yamabe flow for arbitrary initial energy. J. Differ. Geom. **69**(2), 217–278 (2005)
3. Brendle, S.: Convergence of the Yamabe flow in dimension 6 and higher. Invent. Math. **170**(3), 541–576 (2007)
4. Brendle, S., Viaclovsky, J.A.: A variational characterization for $\sigma_{n/2}$. Calc. Var. Partial Differ. Equ. **20**, 399–402 (2004)
5. Catino, G., Mantegazza, C., Mazzieri, L.: On the global structure of conformal gradient solitons with nonnegative Ricci tensor. Commun. Contemp. Math. **14**(06), 1250045 (2012)
6. Chang, S.Y.A., Gursky, M.J., Yang, P.: An a priori estimate for a fully nonlinear equation on four-manifolds. J. Anal. Math. **87**(1), 151–186 (2002)
7. Chow, B.: The Yamabe flow on locally conformally flat manifolds with positive Ricci curvature. Comm. Pure Appl. Math. **45**(8), 1003–1014 (1992)
8. Chow, B., Lu, P., Ni, L.: Hamilton's Ricci Flow, vol. 77. American Mathematical Society/Science Press, Providence (2023)
9. Daskalopoulos, P., Sesum, N.: On the extinction profile of solutions to fast diffusion. J. Reine Angew. Math. **622**, 95–119 (2008)
10. Daskalopoulos, P., Sesum, N.: The classification of locally conformally flat Yamabe solitons. Adv. Math. **240**, 346–369 (2013)
11. Duncan, J.A.J., Wang, Y.: Fully nonlinear Yamabe-type problems on non-compact manifolds. Preprint. arXiv:2401.07808 (2024)
12. Espinal, M.F., Sáez, M.: On the existence and classification of k-Yamabe solitons. J. Geom. Anal. **35**(7), 1–45 (2025)

13. Guan, P., Wang, G.: A fully nonlinear conformal flow on locally conformally flat manifolds. J. Reine Angew. Math. (Crelles Journal) **2003**(557), 219–238 (2003)
14. Hamilton, R.S.: Lectures on geometric flows. Unpublished manuscript, 1 (1989)
15. Jin, Z.R.: A counterexample to the Yamabe problem for complete noncompact manifolds. In: Partial Differential Equations (Tianjin, 1986), volume 1306 of Lecture Notes in Math., pp. 93–101. Springer, Berlin (1988)
16. Li, A., Li, Y.: On some conformally invariant fully nonlinear equations. Comm. Pure Appl. Math. **56**, 1416–1464 (2003)
17. Li, J., Sheng, W.: Deforming metrics with negative curvature by a fully nonlinear flow. Calc. Var. Partial Differ. Equ. **23**(1), 33–50 (2005)
18. Mantegazza, C.: Lecture Notes on Mean Curvature Flow. Progress in Mathematics, vol. 290. Birkhäuser, Basel (2011)
19. Schoen, R.: Conformal deformation of a Riemannian metric to constant scalar curvature. J. Differential Geom. **20**(2), 479–495 (1984)
20. Schoen, R., Yau, S.-T.: Conformally flat manifolds, Kleinian groups and scalar curvature. Invent. Math. **92**(1), 47–71 (1988)
21. Trudinger, N.S.: Remarks concerning the conformal deformation of Riemannian structures on compact manifolds. Ann. Sc. Norm. Super. Pisa Cl. Sci. **22**(2), 265–274 (1968)
22. Viaclovsky, J.A.: Conformal geometry, contact geometry, and the calculus of variations. Duke Math. J. **101**(2), 283 (2000)
23. Yamabe, H.: On a deformation of Riemannian structures on compact manifolds. Osaka J. Math. **12**(1), 21–37 (1960)
24. Ye, R.: Global existence and convergence of Yamabe flow. J. Differ. Geom. **39**(1), 35–50 (1994)

The Continuity Problem of Lyapunov Exponents for Measures with Noncompact Support

Adriana Cristina Sánchez Chavarría

Abstract In this note we address the problem of studying the continuity of the Lyapunov exponents of the Bernoulli shift as functions of a measure with noncompact support.

Keywords Lyapunov exponents · Bernoulli shift · Linear cocycles

1 Introduction

Given an invertible measure preserving transformation $f: (M, \mu) \to (M, \mu)$ of a standard probability space and a measurable function $A: M \to GL(d, \mathbb{R})$ we define the *linear cocycle* defined by A over f as the function

$$F : M \times \mathbb{R}^d \to M \times R^d, \quad F(x, v) = (f(x), A(x)v), \tag{1}$$

with iterates $F^n(x, v) = (f^n(x), A^n(x)v)$ dynamically defined for $n \geq 1$ by

$$A^n(x) = A(f^{n-1}(x)) \cdots A(f(x))A(x).$$

If f is invertible, so is F and $F^{-n}(x, v) = (f^{-n}(x), A^{-n}(x)v)$ with

$$A^{-n}(x) = A(f^{-n}(x))^{-1} \cdots A(f^{-1}(x))^{-1}.$$

The simplest examples of linear cocycles are given by derivative transformations of smooth dynamical systems: the cocycle *generated* by $A(x) = Df(x)$ over f is called the *derivative cocycle*. In the context of hyperbolic dynamical systems, one can understand certain properties of the base dynamics f by examining the behavior

A. C. Sánchez Chavarría (✉)
Universidad de Costa Rica, San José, Costa Rica
e-mail: adriana.sanchez_c@ucr.ac.cr

P. E. Harris et al. (eds.), *Women in Mathematics in Latin America*,
Research Perspectives Ghent Analysis and PDE Center 14,
https://doi.org/10.1007/978-3-031-97857-9_18

of its derivative Df on the tangent space. Similarly, by studying linear cocycles, which offer a more general and flexible framework, we can gain insights into the properties of f. Linear cocycles naturally appear in various contexts, including the spectral theory of Schrödinger operators.

In Ergodic Theory, Lyapunov exponents quantify the average exponential growth rate of the norm of iterates of a cocycle along invariant subspaces within the fibers. One of the key issues in studying Lyapunov exponents of linear cocycles is the regularity problem. This involves examining how the Lyapunov exponents behave in relation to the underlying data, such as the measure and the cocycle itself.

When the base dynamics is a Bernoulli shift, the dependence of Lyapunov exponents on the corresponding probability distribution $\mu = p^{\mathbb{Z}}$ has been explored by several researchers. Furstenberg and Kifer [7] proved that this dependence is continuous at every quasi-irreducible point, that is, such that there is at most one subspace invariant under every matrix in the support of p. Also under an irreducibility condition, LePage [8] demonstrated the Hölder continuity of the Lyapunov exponents.

Bocker and Viana [4] established that continuity holds at every point for compactly supported probability distributions. Building on this, Backes et al. [2] significantly extended the result to linear cocycles with invariant holonomies over hyperbolic systems with local product structures on compact spaces. Additionally, Tall and Viana [10] proved that Hölder continuity is maintained in the compactly supported case when the exponents $\lambda^-(p)$ and $\lambda^+(p)$ are distinct. When these exponents coincide, a weaker but still quantitative form of continuity, known as log-Hölder continuity, is observed. More recently, Avila et al. [1] generalized Bocker and Viana's result for any dimension d, also for measures with compact support.

The aim of this note is to provide a summary of recent research on the continuity problem of Lyapunov exponents for Bernoulli measures in the general case, where the support of the probability measure p is not necessarily compact.

2 Preliminaries

Let $M = SL(2, \mathbb{R})^{\mathbb{Z}}$ and, let $f : M \to M$ be the shift map over M defined by

$$(\alpha_n)_n \mapsto (\alpha_{n+1})_n.$$

Consider the function

$$A : M \to SL(2, \mathbb{R}), \quad (\alpha_n)_n \mapsto \alpha_0,$$

and define its n-th iterate, the *product of random matrices*, by

$$A^n((\alpha_k)_k) = \alpha_{n-1} \cdots \alpha_0.$$

Consider the linear cocycle defined by A over f. Since the base dynamics f is fixed, we allow ourselves to call A itself a cocycle.

Given an invariant measure p in $SL(2,\mathbb{R})$ the associated Bernoulli measure $\mu = p^{\mathbb{Z}}$ is an invariant measure on M. Let $L^1(\mu)$ denote the space of μ-integrable functions on M. Assuming that $\log^+ \|A^{\pm}\| \in L^1(\mu)$ it follows from Furstenberg-Kesten theorem [6, Theorem 2], that

$$\lambda_+(x) = \lim_n \frac{1}{n} \log \|A^n(x)\| \quad \text{and} \quad \lambda_-(x) = \lim_n \frac{1}{n} \log \|A^{-n}(x)\|^{-1},$$

exist for μ-almost every $x \in M$. We call such limits **Lyapunov exponents**.

Furthermore, since the Lyapunov exponents are f-invariant, ergodicity of the Bernoulli measure implies that they are constant for μ-almost every point x. In this case we write $\lambda_+(x) = \lambda_+(\mu)$ and $\lambda_-(x) = \lambda_-(\mu)$.

3 Continuity Problem

Let us assume the measure p has no compact support. As previously mentioned, this work focuses on the continuous dependence of Lyapunov exponents. To address this, we first need to define the primary topology that will be used throughout our analysis.

There is more than one useful topology in the space of probability measures. For example the uniform topology or the pointwise topology, and when M is a metric space one may also consider the weak* topology (see [11]). The *weak* topology* is the smallest topology such that $\nu \mapsto \int \varphi \, d\nu$ is continuous for every bounded continuous function $\varphi : SL(2,\mathbb{R}) \to \mathbb{R}$. In other words, we say that a sequence of measures $(\mu_n)_n$ converges to a measure μ in the weak* topology if and only if

$$\left| \int \psi d\mu_n - \int \psi \mu \right| \to 0,$$

when n goes to infinity, for all ψ bounded continuous functions.

It is well-known that when measures have compact support, Lyapunov exponents are semicontinuous with respect to the weak* topology (see, for example, [11, Chapter 9]. However, this property does not necessarily hold in the non-compact setting. If Lyapunov exponents were semicontinuous in the non-compact case, then every measure with a vanishing Lyapunov exponent would be a point of continuity. The following example taken from [9] demonstrates that this is not the case.

Example 1 Define the function $\alpha : \mathbb{N} \to SL(2, \mathbb{R})$ by

$$\alpha(2k-1) = \begin{pmatrix} \sigma_k & 0 \\ 0 & \sigma_k^{-1} \end{pmatrix}$$

$$\alpha(2k) = \begin{pmatrix} \sigma_k^{-1} & 0 \\ 0 & \sigma_k \end{pmatrix}$$

where $(\sigma_k)_k$ is an increasing sequence such that $\sigma_1 > 1$ and $\sigma_k \to +\infty$.

Let $\mu = q^{\mathbb{Z}}$ be a measure in M where q is the measure on $SL(2, \mathbb{R})$ given by

$$q = \sum_{k \in \mathbb{N}} p_k \delta_{\alpha(k)},$$

with $\sum p_k = 1, 0 < p_k < 1$ for all $k \in \mathbb{N}$.

Consider $0 < r < 1/2 < s < 1$, and $l = s/r > 1$. Let us take $\sigma_k = e^{l^k}$ for all k, which is an increasing sequence provided that $l > 1$. For $k \geq 2$ take $p_{2k-1} = p_{2k} = r^k$. Since $0 < r < 1/2$ it is easy to see that

$$\sum_{k \geq 3} p_k = 2 \sum_{k \geq 2} p_{2k} = 2 \sum_{k \geq 2} r^k = 2 \frac{r^2}{1-r} < 1,$$

and, in order to have a probability measure, it is enough to take

$$p_1 = p_2 = \frac{1}{2} \left(1 - 2 \frac{r^2}{1-r} \right).$$

By an easy computation one can show that $\log \|A\| \in L^1(\mu)$. Moreover, since $p_{2k-1} = p_{2k}$ for all k then $\lambda_+(q) = 0$.

Fix $n_0 > 1$ large enough so $\frac{1}{2}\left(1 - 2\frac{r^2}{1-r}\right) > l^{-n}$ for all $n \geq n_0$, and consider $q_n = \sum_k q_k^n \delta_{\alpha(k)}$ where for $n \geq n_0$

$$q_{2n}^n = l^{-n} + r^n,$$

$$q_2^n = \frac{1}{2} \left(1 - \frac{r^2}{1-r} \right) - l^{-n}$$

$$q_k^n = p_k \text{ other case.}$$

Thus, since $q_k^n \to p_k$ when $n \to \infty$ for all k, it is easy to see that q_n converges in the weak* topology to q.

Notice that

$$\lambda_+(q_n) = (q_{2n-1}^n - q_{2n}^n) \log \sigma_n + (q_1^n - q_2^n) \log \sigma_1 = l^{-n} l^n + l^{-n+1},$$

which is equal to $1 + l^{-n+1} \geq 1$ for all $n \geq n_0$. So $\lambda_+(q_n) \geq 1$ for all $n \geq n_0$ and, therefore, it does not converge to $\lambda_+(q)$.

This example indicates that it is necessary to consider a stronger topology than the weak* topology. For that reason we consider the Wasserstein topology.

Let (M, d) be a Polish space, that is a complete separable metric space. The *Wasserstein space* is the space of probability measures on M with finite moments of order 1, that is,

$$P_1(M) := \left\{ \mu \in P(M) : \int_M d(x_0, x) d\mu(x) < +\infty \right\},$$

where $x_0 \in M$ is arbitrary and $P(M)$ denotes the space of Borel probability measures on M. This does not depend on the choice of the point x_0.

We use $\mu_k \xrightarrow{*} \mu$ to mean convergence in the weak* topology. Moreover, $\mu_k \xrightarrow{W} \mu$ will mean convergence in the Wasserstein topology, characterized by the following (see [12, Definition 6.8]):

$$\mu_k \xrightarrow{*} \mu \quad \text{and} \quad \int d(x_0, x) d\mu_k(x) \to \int d(x_0, x) d\mu(x).$$

The **Wasserstein distance** between two probability measures μ and ν is defined by the formula

$$W_1(\mu, \nu) = \inf_{\pi \in \Pi(\mu, \nu)} \int_M d(x, y) d\pi(x, y), \tag{2}$$

where the infimum is taken over the set $\Pi(\mu, \nu)$ of all the couplings of μ and ν. A *coupling* of μ and ν is a measure π on $M \times N$, such that π projects to μ and ν on the first and second coordinate respectively. This distance metrizes the Wasserstein topology (see [12, Theorem 6.18]).

Let us consider the polish space $SL(2, \mathbb{R})$ with the metric $d(\alpha, \beta) = \|\alpha - \beta\|$. The advantage of using the Wasserstein topology is that all probability measures in $P_1(SL(2, \mathbb{R}))$ have finite moment of order 1, and so their Lyapunov exponents always exist. Moreover, in [9] Sanchez and Viana proved that, in the Wasserstein topology, the function defined on $P_1(SL(2, \mathbb{R}))$ by $p \to \lambda_+(p)$ is upper semi-continuous. The same remains valid for the function $p \to \lambda_-(p)$ with lower semi-continuity.

However, they also proved that, even though the Wasserstein topology is stronger than the weak* topology, it is not stronger enough to guarantee the continuity in the case of measures of non-compact support.

Consider the function $\alpha : \mathbb{N} \to SL(2, \mathbb{R})$ defined by the hyperbolic matrices

$$\alpha(k) = \begin{pmatrix} k & 0 \\ 0 & k^{-1} \end{pmatrix}.$$

Take $m \in \mathbb{N}$ the smallest natural number bigger than 1 such that $\sum_{n \geq m} e^{-\sqrt{n}} < 1$, which exist since $\sum_k e^{-\sqrt{k}}$ is convergent, and define the following weights:

$$p_k = e^{-\sqrt{k}}, \text{ if } k \geq m,$$
$$p_1 = 1 - \sum_{n \geq m} e^{-\sqrt{n}},$$
$$p_k = 0, \text{ otherwise.}$$

It is obvious from the definition that $\sum_k p_k = 1$. Hence, we define the probability measure $q = \sum p_k \delta_{\alpha(k)} \in P_1(SL(2, \mathbb{R}))$. Moreover, it is easy to see that $\log \|A\| \in L^1(\mu)$ and, $\lambda_+(p) = \sum_{k=m}^{\infty} e^{-\sqrt{k}} \log k > 0$.

Now, consider the rotation matrix $B = \begin{pmatrix} 0 & -1 \\ 1 & 0 \end{pmatrix}$ and for each n consider

$$\beta_n(k) = \begin{cases} \alpha(k) & \text{if } k \neq n, \\ B & \text{if } k = n. \end{cases}$$

Define the probability measures $q_n = \sum_k p_k \delta_{\beta_n(k)}$. In a similar manner to the previous discussion, we can observe that for all n, these measures belong to $P_1(SL(2, \mathbb{R}))$ and $\log \|A\| \in L^1(q_n)$. Moreover, since

$$W(q_n, p) \leq p_n d(\alpha(n), B) \sim n e^{-\sqrt{n}}$$

which goes to 0 if n goes to ∞, q_n converges to q in the Wasserstein topology.

Assume there exist N such that $\lambda_+(q_N) > 0 > \lambda_-(q_N)$. We will consider the distance in the projective space $\mathbb{P}$ given by

$$\delta([v], [w]) := \frac{\|v \wedge w\|}{\|v\| \|w\|} = \sin(\angle(v, w)).$$

Consider the family $\mathcal{F} = \{V, H\}$, where $V = [e_2]$ and $H = [e_1]$ represent the vertical and horizontal axes, respectively. By the definition of the measure q_N, it is evident that this family is invariant under every matrix in the support of this measure. Furthermore, if $x = (x_k)_k \in M$, then we can observe that for every m

$$\delta(A^m(x)H, A^m(x)V) \geq \delta(H, V) = 1. \tag{3}$$

On the other hand, since the cocycle A takes values in $SL(2, \mathbb{R})$, we have $\| \wedge^2 A^m(x)\| = 1$ and therefore, for every unit vectors v and w

$$\|A^m(x)v \wedge A^m(x)w\| \leq \|v \wedge w\| \leq \|v\| \|w\| \leq 1.$$

It is widely known that, for q_N-almost every $x \in M$

$$\lambda_+(q_N) + \lambda_-(q_N) = \lim_m \frac{1}{m} \log \| \wedge^2 A^m(x) \|.$$

Note that q_N is irreducible, meaning that there is no proper subspace of $\mathbb{R}^2$ that is invariant under all the matrices in the support of q_N. Consequently, we have

$$\lambda_+(q_N) = \lim_m \frac{1}{m} \log \|A^m(x)w\|,$$

for every unit vector w. For a deeper discussion of the two results mentioned above we refer the reader to [5, Chapter III].

Thus, we have for every unit vectors v and w

$$\lim_m \frac{1}{m} \log \frac{\| \wedge^2 A^m(x)\|}{\|A^m(x)v\| \|A^m(x)w\|} = \lambda_-(q_N) - \lambda_+(q_N) = -2\lambda_+(q_N) < 0.$$

Therefore, using the fact that for any positive numerical sequence $a_n > 0$ satisfying $\lim_{n\to\infty} \log a_n < 0$ it follows that $lim_{n\to\infty} a_n = 0$, we can conclude that

$$\lim_m \delta(A^m(x)H, A^m(x)V) \leq \lim_m \frac{\| \wedge^2 A^m(x)\|}{\|A^m(x)e_1\| \|A^m(x)e_2\|} = 0,$$

which is a contradiction with (3).

This example illustrates that the Wasserstein topology alone is insufficient to guarantee the continuity of the Lyapunov exponents. The primary issue is that the supports of the measures q_n diverge from the support of q. This suggests that we need an additional condition to ensure the "convergence" of the supports. Bocker and Viana [4] addressed a similar issue for measures with compact support by incorporating such an assumption. However, in the non-compact setting, we also need to account for the behavior of the iterates of the cocycles. Therefore, we are left with the following open problem:

Open Problem

Find a criterion for the convergence of the supports of measures such that, in conjunction with the Wasserstein topology, allows us to ensure the continuity of the Lyapunov exponents by controlling the behavior of the iterates of the cocycles.

4 Future Work

While this study provides valuable insights into the continuity problem of Lyapunov exponents with measures of non-compact support, several areas warrant further exploration. One promising direction for future research is to investigate the conditions for higher regularity conditions in different base dynamics or higher dimensions.

Tall and Viana [10] demonstrated that for random $GL(2, \mathbb{R})$ cocycles, the Lyapunov exponents are pointwise Hölder continuous when the exponents are distinct. A promising avenue for future research is to generalize this result to higher dimensions, leveraging the findings presented in Avila et al. [1].

Alternatively, one can focus on the continuity or regularity problem for different type of cocycles. Bezerra et al. [3] addressed the problem of a random product of quasi-periodic cocycles depending real analytically on the transition probabilities p whenever $\lambda_+(p)$ is simple for finitely supported measures. So a natural question is to explore the case of general measures.

In conclusion, there are several research directions to explore in the study of the continuity of Lyapunov exponents, each presenting distinct challenges. Addressing these research gaps can lead to the development of new techniques and deepen our understanding of how Lyapunov exponents behave in relation to the underlying data.

References

1. Avila, A., Eskin, A., Viana, M.: Continuity of the Lyapunov exponents of random matrix products. (2023). arXiv_preprint_arXiv:2305.06009
2. Backes, L., Brown, A., Butler, C.: Continuity of Lyapunov exponents for cocycles with invariant holonomies. J. Mod. Dyn. **12**, 261–283 (2018)
3. Bezerra, J., Sánchez, A., Tall, E.H.Y.: Analyticity of the Lyapunov exponents of random products of quasi-periodic cocycles. Nonlinearity **36**(6), 3467–3482 (2023)
4. Bocker, C., Viana, M.: Continuity of Lyapunov exponents for random two-dimensional matrices. Ergodic Theory Dyn. Syst. **37**, 1413–1442 (2017)
5. Bougerol, P., et al.: Products of Random Matrices with Applications to Schrödinger Operators, vol. 8. Springer Science & Business Media, New York (2012)
6. Furstenberg, H., Kesten, H.: Products of random matrices. Ann. Math. Stat. **31**, 457–469 (1960)
7. Furstenberg, H., Kifer, Y.: Random matrix products and measures in projective spaces. Ann. Inst. H. Poincaré Probab. Stat. **10**, 12–32 (1983)
8. Le Page, É.: Régularité du plus grand exposant caractéristique des produits de matrices aléatoires indépendantes et applications. Israel J. Math. **25**, 109–142 (1989)
9. Sánchez, A., Viana, M.: Lyapunov exponents of probability distributions with non-compact support (2020). https://arxiv.org/pdf/1810.03061.pdf
10. Tall, E.Y., Viana, M.: Moduli of continuity of Lyapunov exponents for random GL(2)-cocycles. Trans. Amer. Math. Soc. **373**, 1343–1383 (2020)
11. Viana, M.: Lectures on Lyapunov Exponents, vol. 145. Cambridge University Press, Cambridge (2014)
12. Villani, C.: Optimal Transport: Old and New, vol. 338. Springer Science & Business Media, New York (2008)

Degree Conditions for Trees in Undirected and Directed Graphs

Maya Stein

Abstract We survey some recent results on bounds on the minimum and maximum degree of graph that ensure it contains all trees of a fixed size, and corresponding results for digraphs. We also present a number of conjectures and open questions.

Keywords Degree conditions · Trees · Digraphs · Extremal graph theory

1 Introduction

At the core of extremal graph theory are subgraph containment problems, in particular those where the existence of a certain subgraph is connected to a condition on the minimum degree of the host graph. We focus on the case when the subgraph in question is a tree. One of the best-known results in this direction is the Komlós–Sárközy–Szemerédi theorem [12] which gives a lower bound on the minimum degree $\delta(G)$ of a graph G that ensures G contains any spanning tree of bounded degree. In recent years, some versions for smaller trees and corresponding lower bounds on $\delta(G)$ have been proposed. We will see below that in these versions, it is necessary to add an additional condition on the host graph, such as a lower bound on the maximum degree $\Delta(G)$ of G. We give a quick overview of the state of the art for graphs in Sect. 2.

A natural question is how to generalise problems of the above type to digraphs, following the increasingly popular trend of extending results from extremal graph theory to the digraph setting. As every graph can be viewed as a digraph, but the

The author acknowledges support by FONDECYT Regular Grant 1221905 and by ANID Basal Grant CMM FB210005.

M. Stein (✉)
University of Chile, Santiago, Chile
e-mail: mstein@dim.uchile.cl

P. E. Harris et al. (eds.), *Women in Mathematics in Latin America*,
Research Perspectives Ghent Analysis and PDE Center 14,
https://doi.org/10.1007/978-3-031-97857-9_19

universe of digraphs is much richer than the world of graphs, this leads to fascinating new challenges. A few new results on using semidegree conditions to find oriented trees in digraphs are surveyed in Sect. 3.

For more context, we refer the interested reader to the surveys [18, 19], and remark that in particular, Sect. 3 here can be seen as a continuation of Sect. 6 of [19]. We remark that for better exposition, some of the results are not stated here in the same generality as they appear in the respective sources.

2 Trees in Graphs

We start by noting that every graph G of minimum degree $\delta(G) \geq k$ contains all trees with k edges, as we can just embed the tree greedily into G. The bound on $\delta(G)$ could not be weaker: Consider the union of several disjoint copies of K_k which gives a graph of minimum degree $k-1$ that does not contain any tree with k edges. This example can be eliminated if we decide to add the constraint that G is connected and sufficiently large. Dirac and independently Erdős and Gallai observed the following.

Proposition 1 (Dirac/Erdős and Gallai (See [4])) *If a graph G has minimum degree at least $k/2$ and a connected component on at least $k+1$ vertices then G contains a path with k edges.*

However, connectivity and large order alone are not helpful if we are looking for trees other than paths (see [18] for details).

Trying to overcome the example from above from a different angle, we may observe that the construction only works if $k < |V(G)|/2$. So for trees that are comparatively large with respect to the host graph G, a better bound on $\delta(G)$ would be possible. This has been confirmed for trees of bounded maximum degree in the following classical result:

Theorem 2 (Komlós, Sárközy and Szemerédi [12]) *For all $\delta > 0$, there are n_0 and c such that every graph G on $n \geq n_0$ vertices with $\delta(G) \geq (1+\delta)\frac{n}{2}$ contains each n-vertex tree T with $\Delta(T) \leq c\frac{n}{\log n}$.*

All bounds in Theorem 2 are tight up to the constant c. In particular, it is necessary to bound $\Delta(T)$. If we wish to omit the bound on $\Delta(T)$, that is, if we wish to find *all* spanning trees, then it will be necessary to require a vertex of large degree in G (because T could be a star). It turns out that having just one vertex of degree at least k is indeed sufficient, if we also elevate the bound on $\delta(G)$:

Theorem 3 (Reed and Stein [14, 15]) *There is n_0 such that for every $n \geq n_0$, every n-vertex graph with $\delta(G) \geq \lfloor 2(n-1)/3 \rfloor$ and $\Delta(G) \geq n-1$ contains each n-vertex tree.*

Theorem 3 solves a special case of the next conjecture, Conjecture 4, which considers smaller trees and corresponding weaker bounds on $\delta(G)$.

Conjecture 4 (Havet et al. [5]) *Every graph of minimum degree at least $\lfloor 2k/3 \rfloor$ and maximum degree at least k contains each k-edge tree.* □

Note that the bounds in Conjecture 4 are tight, which can be seen by considering the following example: Assume 3 divides k and consider the tree obtained from identifying the starting vertices of three distinct $\frac{k}{3}$-edge paths. This tree is not a subgraph of the graph obtained by taking two cliques of size $2k/3 - 1$ and adding a *universal vertex* (i.e., a vertex adjacent to all other vertices).

As a cross-over of Conjecture 4 and Theorem 2, it was conjectured in [2] that every graph of minimum degree at least $k/2$ and maximum degree at least $2k$ contains each k-edge tree. However, although an approximate version of this statement holds for large bounded degree trees [2], it is not true in general. Hyde and Reed [6] recently gave a counterexample. The following may still hold:

Conjecture 5 *Is it true that every graph with $\delta(G) > k/2$ and with $\Delta(G) \geq 2k$ contains each k-edge tree?* □

Asymptotic versions of Conjectures 4 and 5 hold for bounded degree trees.

Theorem 6 (Besomi et al. [2]) *For every $\eta > 0$ there is $n_0 \in \mathbb{N}$ such that for all $n \geq n_0$ and all $k \geq \eta n$ every n-vertex graph G fulfilling*

(a) $\delta(G) \geq (1+\eta)k/2$ and $\Delta(G) \geq 2(1+\eta)k$; or
(b) $\delta(G) \geq (1+\eta)2k/3$ and $\Delta(G) \geq (1+\eta)k$

contains every k-edge tree T with $\Delta(T) \leq k^{1/67}$ as a subgraph.

Also, Hyde and Reed [6] recently proved the following for all host graphs.

Theorem 7 (Hyde and Reed [6]) *There is a function $f : \mathbb{N} \to \mathbb{N}$ such that every graph G with $\delta(G) \geq k/2$ and $\Delta(G) \geq f(k)$ contains each k-edge tree.*

Earlier it had been shown by Havet et al. [5] that there is an absolute constant $\gamma > 0$ such that every graph G with $\delta(G) \geq (1-\gamma)k$ and $\Delta(G) \geq k$ contains each k-edge tree.

If we wish to leave the minimum degree condition at $k/2$ but avoid requiring the existence of a vertex of very large degree, then a natural candidate for an alternative additional condition is to ask for some kind of expansion of G. In particular, forbidding short cycles should be helpful for finding trees. Note that it will still be necessary to require a vertex of degree at least k, because of the stars. The following result shows that it suffices to forbid 4-cycles.

Theorem 8 (Saclé and Woźniak (see Sect. 4 of [16])) *Every graph G having no 4-cycles and fulfilling $\delta(G) \geq k/2$ and $\Delta(G) \geq k$ contains any k-edge tree.*

Assuming G has large girth (where the *girth* of a graph G is the largest g such that G has no cycles of length $< g$), the bound on the minimum degree can be even smaller, as long as T has somewhat bounded degree.

Theorem 9 (Jiang [8]) *Every graph G of girth at least $2\ell + 1$ with $\delta(G) \geq \max\{k/\ell, \Delta(T)\}$ contains any k-edge tree.*

Note that the additional condition $\delta(G) \geq \Delta(T)$ is indeed necessary, which can be seen by considering a balanced double star and a suitable host graph.

3 Oriented Trees in Digraphs

We consider both *digraphs*, which may have up to one edge in each direction between any pair of vertices, and *oriented graphs* which have at most one edge between any pair of vertices. Loops are forbidden in both cases. Since the edges of a digraph have directions, there are several possibilities when it comes to translating the notion of the minimum degree to a digraph. The most popular one is the *minimum semidegree* $\delta^0(D)$, which is defined as the minimum over all the indegrees and all the outdegrees of the vertices of the digraph D.

In analogy to the graph case, every digraph of minimum semidegree at least k contains every oriented tree T with k edges. The bound cannot be weakened since T may be an oriented star. Even if T is an oriented path, a bound below k will not be sufficient: to see this, we can consider the disjoint union of *complete digraphs* (i.e. digraphs having all possible edges) of order k, which has minimum semidegree $k-1$ but contains no oriented subgraph with k edges.

However, if we are only looking for oriented paths in oriented graphs, a generalisation of Proposition 1 may be possible. Indeed, the following was conjectured in [18].

Conjecture 10 ([18]) *Every oriented graph of minimum semidegree exceeding $k/2$ contains any oriented path of length k.* □

Results of Kelly [10] imply that this is true if k is larger than $3n/4 + o(n)$, and an even earlier result of Jackson [7] shows that the conjecture holds for *directed paths*, i.e. oriented paths where each vertex has in- and outdegree at most one. Some more evidence can be found in [3, 11, 17].

One nice property of paths, apart from not having vertices of high degree, is that if they have an even number of vertices, they are *balanced*, i.e. their partition classes have the same size. It may be possible to extend Conjecture 10 to balanced oriented trees. In this direction, it was conjectured in [19] that every oriented n-vertex graph D with $\delta^0(D) > k/2$ contains each balanced antidirected k-edge tree with $\Delta(T) \leq o(n)$, where $\Delta(T)$ denotes *the maximum total degree of T*, i.e. the maximum degree of the underlying undirected tree, and a tree is *antidirected*, if each of its vertices has either outdegree or indegree 0. With a weaker bound on $\Delta(T)$, this conjecture is true if D is large, and k is large compared to the order of D:

Theorem 11 (Stein and Zárate-Guerén [21]) *For all $\eta > 0$, $c \in \mathbb{N}$ there is n_0 such that for all $n \geq n_0$ and $k \geq \eta n$, every oriented graph D on n vertices with $\delta^0(D) > (1+\eta)k/2$ contains every balanced antidirected tree T with k edges and with $\Delta(T) \leq (\log(n))^c$.*

We remark that the condition that T is balanced is necessary. To see this, assume k is even and consider the oriented graph D on sets V_1, V_2, V_3, each of size $k/2$, having all edges from V_i to V_{i+1} (mod 3). Then $\delta^0(D) = k/2$. Let T be an unbalanced antidirected tree. It is easy to see that T is not contained in D. In fact, the more unbalanced T is, the larger we could make the sets V_i, and D would still not contain T. The extreme case is when T is an antidirected star and each V_i has size $k-2$.

In the case that T and D have the same order and $\Delta(T)$ is bounded by a constant, we can omit the condition that T is antidirected in Theorem 11, and lower the bound on $\delta^0(D)$ to about $3n/8$, as we will see in the following new result. Interestingly, this bound on the minimum semidegree corresponds to the bound for oriented Hamilton cycles in oriented graphs due to Kelly [10].

Theorem 12 (Araújo et al. [1]) *For every $\eta > 0$ and $\Delta \in \mathbb{N}$, there is an integer n_0 such that for all $n \geq n_0$ each oriented graph D on n vertices with $\delta^0(D) \geq (3/8 + \eta)n$ contains each oriented tree T on n vertices with $\Delta(T) \leq \Delta$.*

Note that Theorem 12 can be viewed as a version of Theorem 2 for oriented graphs (albeit with a worse bound on $\Delta(T)$). Some years ago, Kathapurkar and Montgomery gave a generalisation of the same theorem to digraphs [9]. Again, all bounds are best possible up to the constant.

Theorem 13 (Kathapurkar and Montgomery [9]) *For each $\eta > 0$, there are $c > 0$ and $n_0 \in \mathbb{N}$ such that every n-vertex digraph on $n \geq n_0$ vertices and with $\delta^0(G) \geq (1/2 + \eta)n$ contains each n-vertex oriented tree with $\Delta(T) \leq cn/\log n$.*

From the examples given above, we know that Theorem 13 becomes false if we change the size of the tree to arbitrary k (independent of n), and accordingly require $\delta^0(D) \geq (1/2 + \eta)k$. However, this changes if we add the restriction that D has a vertex of large outdegree and a vertex of large indegree. Let us define $\Delta^{\pm}(D)$ as the largest ℓ such that D has a vertex of outdegree $\geq \ell$ and a vertex of indegree $\geq \ell$ (these two vertices may coincide).

In the spirit of Conjectures 4 and 5 we have the following new result for balanced antidirected trees.

Theorem 14 (Kontogeorgiou et al. [13]) *For all $\eta > 0$ and $c \in \mathbb{N}$, there exists $n_0 \in \mathbb{N}$ such that for all $n \geq n_0$ and $k \geq \eta n$, every digraph D on n vertices fulfilling*

(a) $\delta^0(D) \geq (\frac{3}{5} + \eta)k$ and $\Delta^{\pm}(D) \geq 2(1 + \eta)k$ or
(b) $\delta^0(D) \geq (\frac{2}{3} + \eta)k$ and $\Delta^{\pm}(D) \geq (1 + \eta)k$

contains every balanced antidirected tree T with k edges and with $\Delta(T) \leq (\log n)^c$.

Another new result is the generalisation of Theorem 8 to digraphs. We now forbid all orientations of the 4-cycle.

Theorem 15 (Stein and Trujillo-Negrete [20]) *Every digraph D having no oriented 4-cycles with $\delta^0(D) \geq k/2$ and $\Delta^{\pm}(D) \geq k$ contains all oriented trees T with k edges.*

It would be interesting to extend also Theorem 9 to digraphs. In this respect, we pose the following question.

Question 16 *Does every oriented graph D of girth at least $2\ell + 1$ with $\delta^0(D) \geq \max\{k/\ell, \Delta(T)\}$ contain every orientation of each k-edge tree?*

References

1. Araújo, P., Santos, G., Stein, M.: Spanning trees in dense oriented graphs, In preparation
2. Besomi, G., Pavez-Signé, M., Stein, M.: Degree conditions for embedding trees. SIAM J. Discrete Math. **33**, 1521–1555 (2019)
3. Chen, B., Hou, X., Zhou, H.: Long antipaths and anticycles in oriented graphs. Preprint. arXiv:2401.05205 (2024)
4. Erdős, P., Gallai, T.: On maximal paths and circuits of graphs. Acta Math. Acad. Sci. Hungar. **10**, 337–356 (1959)
5. Havet, F., Reed, B., Stein, M., Wood, D.: A variant of the Erdős-Sós conjecture. J. Graph Theory **94**(1), 131–158 (2020)
6. Hyde, J., Reed, B.: Graphs of minimum degree at least d/2 and large enough maximum degree embed every tree with d vertices. Procedia Comput. Sci. **223**, 217–222 (2023), XII Latin-American Algorithms, Graphs and Optimization Symposium (LAGOS 2023)
7. Jackson, B.: Long paths and cycles in oriented graphs. J. Graph Theory **5**(2), 145–157 (1981)
8. Jiang, T.: On a conjecture about trees in graphs with large girth. J. Combin. Theory Ser. B **83**(2), 221–232 (2001)
9. Kathapurkar, A., Montgomery, R.: Spanning trees in dense directed graphs. J. Combin. Theory Ser. B **156**, 223–249 (2022)
10. Kelly, L.: Arbitrary orientations of hamilton cycles in oriented graphs. Electron. J. Combin. **18**, 44 (2011)
11. Klimošová, T., Stein, M.: Antipaths in oriented graphs. Discrete Math. **346**, 113515 (2023)
12. Komlós, J., Sárközy, G.N., Szemerédi, E.: Spanning trees in dense graphs. Combin. Probab. Comput. **10**(5), 397–416 (2001)
13. Kontogeorgiou, G., Santos, G., Stein, M.: Antidirected trees in directed graphs. Preprint 2025, arXiv:2501.11726
14. Reed, B., Stein, M.: Spanning trees in graphs of high minimum degree with a universal vertex I: an asymptotic result. J. Graph Theory **102**(4), 737–783 (2023)
15. Reed, B., Stein, M.: Spanning trees in graphs of high minimum degree with a universal vertex II: a tight result. J. Graph Theory **102**(4), 797–821 (2023)
16. Saclé, J.-F., Woźniak, M.: A note on the Erdős–Sós conjecture for graphs without C_4. J. Combin. Theory (Ser. B) **70**(2), 229–234 (1997)
17. Skokan, J., Tyomkyn, M.: Alternating paths in oriented graphs with large semidegree. Preprint. arXiv:2406.03166 (2024)
18. Stein, M.: Tree containment and degree conditions. In: Raigorodskii, A.M., Rassias, M.Th. (eds.) Discrete Mathematics and Applications, pp. 459–486. Springer International Publishing, Cham (2020)
19. Stein, M.: Oriented trees and paths in digraphs. In: Fischer, F., Johnson, R. (eds.) Surveys in Combinatorics 2024. London Mathematical Society Lecture Note Series, pp. 271–295. Cambridge University Press, Cambridge (2024)

20. Stein, M., Trujillo-Negrete, A.: Oriented trees in digraphs without oriented 4-cycles. Preprint. arXiv:2411.13483 (2024)
21. Stein, M., Zárate-Guerén, C.: Antidirected subgraphs of oriented graphs. Combin. Probab. Comput. **33**(4), 1–21 (2024)

The manufacturer's authorised representative in the EU is Springer Nature Customer Service Centre GmbH, Europaplatz 3, 69115 Heidelberg, Germany. If you have any concerns regarding our products, please contact ProductSafety@springernature.com

Printed and bound by CPI Group (UK) Ltd, Croydon, CR0 4YY

17/09/2026

02218458-0004